Kai Bester

**Personal Care Compounds
in the Environment**

1807–2007 Knowledge for Generations

Each generation has its unique needs and aspirations. When Charles Wiley first opened his small printing shop in lower Manhattan in 1807, it was a generation of boundless potential searching for an identity. And we were there, helping to define a new American literary tradition. Over half a century later, in the midst of the Second Industrial Revolution, it was a generation focused on building the future. Once again, we were there, supplying the critical scientific, technical, and engineering knowledge that helped frame the world. Throughout the 20th Century, and into the new millennium, nations began to reach out beyond their own borders and a new international community was born. Wiley was there, expanding its operations around the world to enable a global exchange of ideas, opinions, and know-how.

For 200 years, Wiley has been an integral part of each generation's journey, enabling the flow of information and understanding necessary to meet their needs and fulfill their aspirations. Today, bold new technologies are changing the way we live and learn. Wiley will be there, providing you the must-have knowledge you need to imagine new worlds, new possibilities, and new opportunities.

Generations come and go, but you can always count on Wiley to provide you the knowledge you need, when and where you need it!

William J. Pesce
President and Chief Executive Officer

Peter Booth Wiley
Chairman of the Board

Kai Bester

Personal Care Compounds in the Environment

Pathways, Fate and Methods for Determination

With Contributions of Stefan Weigel,
Michael P. Schlüsener and Jens A. Andresen

WILEY-VCH Verlag GmbH & Co. KGaA

The Author

Dr. Kai Bester
Institut of Environmental Analytical
Chemistry
Duisburg-Essen
Universitätsstrasse 15
45114 Essen
Germany

■ All books published by Wiley-VCH are carefully
produced. Nevertheless, authors, editors, and
publisher do not warrant the information contained
in these books, including this book, to be free of
errors. Readers are advised to keep in mind that
statements, data, illustrations, procedural details or
other items may inadvertently be inaccurate.

Library of Congress Card No.: applied for

British Library Cataloguing-in-Publication Data
A catalogue record for this book is available
from the British Library.

**Bibliographic information published by
the Deutsche Nationalbibliothek**
The Deutsche Nationalbibliothek lists this publica-
tion in the Deutsche Nationalbibliografie; detailed
bibliographic data are available in the Internet at
http://dnb.d-nb.de.

© 2007 WILEY-VCH Verlag GmbH & Co. KGaA,
Weinheim

Typesetting   K+V Fotosatz GmbH, Beerfelden
Printing   Strauss GmbH, Mörfelden
Bookbinding   Litges & Dopf Buchbinderei
GmbH, Heppenheim

Printed in the Federal Republic of Germany
Printed on acid-free paper

ISBN   978-3-527-31567-3

Contents

Preface *IX*

Acknowledgments *XI*

List of Contributors *XIII*

List of Abbreviations *XV*

1 **Introduction** *1*
1.1 General Considerations (*Kai Bester*) *1*
1.2 Introduction to Sewage Treatment Plant Functions *2*
1.3 Enantioselective Analysis in Environmental Research *4*
1.3.1 Enantioselective Gas Chromatography Techniques *4*
1.3.1.1 Applications of Enantioselective Gas Chromatography *6*
1.3.1.2 New Developments *7*
1.3.2 Enantioselective HPLC *7*
1.3.2.1 Applications of Enantioselective HPLC *7*

2 **Environmental Studies: Sources and Pathways** *9*
2.1 Synthetic Fragrance Compounds in the Environment (*Kai Bester*) *9*
2.1.1 Polycyclic Musk Fragrances in Sewage Treatment Plants *10*
2.1.1.1 Experimental Background *10*
2.1.1.2 Mass Balance Assessment *13*
2.1.1.3 Multi-step Process Study on Polycyclic Musks *18*
2.1.2 Polycyclic Musk Fragrances in Diverse Sludge Samples *23*
2.1.3 Polycyclic Musk Fragrances in Surface Waters *24*
2.1.3.1 Experimental Methods *25*
2.1.3.2 Results and Discussion *29*
2.1.4 Polycyclic Musk Fragrances in the North Sea *38*
2.1.5 OTNE and Other Fragrances in the Environment *44*
2.1.5.1 Methods *45*
2.1.5.2 Results and Discussion *46*
2.1.6 Other Fragrances: Nitroaromatic Musks and Macrocyclic Musks *50*

Personal Care Compounds in the Environment: Pathways, Fate, and Methods for Determination. Kai Bester
Copyright © 2007 WILEY-VCH Verlag GmbH & Co. KGaA, Weinheim
ISBN: 978-3-527-31567-3

2.1.7   Behavior of Polycyclic and Other Musk Fragrances
            in the Environment   *53*
2.2   The Bactericide Triclosan and Its Transformation Product Methyl
            Triclosan in the Aquatic Environment (*Kai Bester*)   *54*
2.2.1   Bactericides from Personal Care Products in Sewage Treatment
            Plants   *54*
2.2.1.1   Materials and Methods   *55*
2.2.1.2   Triclosan Balances in a Sewage Treatment Plant   *57*
2.2.1.3   Triclosan in Multi-step Processes in Sewage Treatment Plants   *59*
2.2.2   Triclosan in Sewage Sludge   *62*
2.2.3   Triclosan in Surface Waters   *63*
2.2.3.1   Estimation of Elimination Constants for Triclosan in a River   *67*
2.2.4   Discussion on Triclosan and Methyl Triclosan
            in the Environment   *69*
2.3   UV Filters/Sunscreens (*Kai Bester*)   *69*
2.3.1   Endocrine Properties of UV Filters   *70*
2.3.2   UV Filters in Aquatic Ecosystems   *72*
2.3.3   Enantioselective Considerations for UV Filters   *73*
2.4   Organophosphate Flame-retardants and Plasticizers
            (*Jens A. Andresen, Stefan Weigel and Kai Bester*)   *74*
2.4.1   Introduction   *74*
2.4.1.1   Flame-retardants   *74*
2.4.1.2   Organophosphate Plasticizers   *76*
2.4.2   The Organophosphate Flame-retardant TCPP
            in a Sewage Treatment Plant   *76*
2.4.2.1   Materials and Methods   *77*
2.4.2.2   Mass Balance Assessment for TCPP
            in a Sewage Treatment Plant   *79*
2.4.2.3   TCPP in Sludge Monitoring   *82*
2.4.2.4   Evaluation of the TCPP Data   *83*
2.4.3   Organophosphate Flame-retardants and Plasticizers
            in Multi-step Sewage Treatment   *83*
2.4.3.1   Materials and Methods   *84*
2.4.3.2   Results and Discussion   *86*
2.4.3.3   Conclusions   *93*
2.4.4   Organophosphorus Flame-retardants and Plasticizers
            in Surface Waters   *93*
2.4.4.1   Materials and Methods   *93*
2.4.4.2   Results and Discussion   *95*
2.4.5   Organophosphates in Drinking Water Purification   *101*
2.4.5.1   Materials and Methods   *102*
2.4.5.2   Results   *105*
2.4.5.3   Conclusions   *112*
2.4.6   Organophosphates and Other Compounds in the North Sea
            and Lake Ontario: A Comparison   *113*

2.4.6.1   Materials and Methods   *115*
2.4.6.2   Results and Discussion   *119*
2.4.6.3   Conclusions   *126*
2.4.7   Overall Discussion on Chlorinated Organophosphorus
           Flame-retardants and Other Compounds   *128*
2.5   Endocrine-disrupting Agents
           (*Michael P. Schlüsener and Kai Bester*)   *128*
2.5.1   Introduction to Endocrine-disrupting Effects   *128*
2.5.2   Estrogenic Hormones and Antibiotics in Wastewater Treatment
           Plants   *136*
2.5.2.1   Description of the Sample Sites   *136*
2.5.2.2   Results and Discussion   *139*
2.5.2.3   Conclusions   *153*
2.5.3   Nonylphenol and Other Compounds in the North Sea   *153*
2.5.3.1   Materials and Methods   *155*
2.5.3.2   Results   *158*
2.5.3.3   Discussion   *161*
2.5.3.4   Conclusions   *163*
2.6   Diverse Compounds (*Kai Bester*)   *164*
2.6.1   Benzothiazoles in Marine Ecosystems   *164*
2.6.1.1   Materials and Methods   *165*
2.6.1.2   Results   *165*
2.6.1.3   Discussion and Conclusions   *171*
2.6.2   Enantioselective Degradation of Bromocyclene in Sewage
           Treatment Plants   *172*
2.6.2.1   Introduction   *172*
2.6.2.2   Methods and Materials   *172*
2.6.2.3   Results and Discussion   *175*

3   **Analytical Chemistry Methods**   *177*
3.1   Fresh and Wastewater (*Kai Bester*)   *177*
3.1.1   Lipophilic Compounds from Fresh and Wastewater
           (GC Analysis)   *177*
3.1.1.1   Sampling   *177*
3.1.1.2   Extractions   *178*
3.1.2   Steroid Hormones, Their Adducts, and Macrolide Antibiotics
           from Wastewater (HPLC-MS/MS Analysis)
           (*Michael P. Schlüsener and Kai Bester*)   *179*
3.1.2.1   Introduction   *179*
3.1.2.2   Experimental Methods   *180*
3.1.2.3   Results and Discussion   *185*
3.1.2.4   Conclusions   *192*
3.2   Seawater   *194*
3.2.1   Lipophilic Compounds in Marine Water Samples (*Kai Bester*)   *194*

3.2.2   Hydrophilic Compounds in Marine Water Samples
            (*Stefan Weigel and Kai Bester*)   *195*
3.2.2.1   Experimental Methods   *197*
3.2.2.2   Results and Discussion   *201*
3.2.2.3   Conclusions   *204*
3.3   Sewage Sludges (*Jens A. Andresen and Kai Bester*)   *205*

4   **Discussion** (*Kai Bester*)   *207*
4.1   Sewage Treatment Plants   *207*
4.2   Limnic Samples   *210*
4.3   Marine Samples   *211*
4.4   Conclusions   *214*

5   **Summary** (*Kai Bester*)   *215*
5.1   Polycyclic Musks AHTN, HHCB, HHCB-lactone, and OTNE   *215*
5.2   Flame-retardants   *218*
5.3   Endocrine Disrupters   *219*
5.4   Triclosan and Methyl Triclosan   *220*

6   **References**   *221*

Subject Index   *241*

Preface

Since the end of the 1960s the general public as well as administrators have been aware that chemical compounds such as pesticides can cause risks. Special awareness was brought to the issue of pesticides and dioxins by Rachel Carson's book *Silent Spring* [1]. The majority of compounds addressed in this work were either of high acute toxicity or carcinogenic. Altered population dynamics and changed fertility were introduced as well, but the impact of these issues was not foreseen at that time. The focus on environmental issues has broadened greatly since that time. Issues of endocrine disruption and long-term (chronic) toxicology as well as ecotoxicology or ecosystem toxicology emerged more clearly in the 1980s and 1990s. The general public became aware of these findings mainly through the book *Our Stolen Future* by Theo Colborn et al. [2]. Because of the pressing issue of long-term (chronic) toxicity and the impossibility of re-capturing chemicals once emitted into the environment, administrators in several countries adopted the so-called precautionary principle, which became especially relevant for large-scale ecosystems such as the North Sea and the Atlantic. The compounds of interest changed from "pollutants" (with proven adverse effects to man or animals) to "xenobiotics." Xenobiotics are manmade chemicals that are used in a multitude of processes. They include compounds used in the technosphere, e.g., additives to concrete such as tributyl phosphates or flame-retardants such as tris-(2-chloro-methyl-ethyl)-phosphate, and the endocrine-disrupting nonylphenols, which are mostly used as plasticizers in epoxy resins or as the ethoxylate derivatives of these nonylphenols. These compounds also are used as industrial detergents in textile production. On the other hand, there are compounds that most of us experience as positive, such as fragrances in washing powders or shampoos, which everyone may use in everyday life. The same holds true for bactericides such as triclosan, which is used as a household bactericide in toothpaste, sportswear, etc. Additionally, there are medicinal compounds that we have become accustomed to using in cases of serious illnesses, e.g., antibiotics, or to cope with lifestyle issues such as a simple hangover, e.g., acetylsalicylic acid (aspirin). It may be appalling to learn that a multitude of antibiotic compounds are used in industrial agriculture, e.g., as growth promoters in pig or cattle fattening, as well.

We use most of these substances to make our life more comfortable or more secure. Though each application may be discussed for its effectiveness and overall use, we should certainly be aware that these substances do not simply vanish

Personal Care Compounds in the Environment: Pathways, Fate, and Methods for Determination. Kai Bester
Copyright © 2007 WILEY-VCH Verlag GmbH & Co. KGaA, Weinheim
ISBN: 978-3-527-31567-3

down the drain but instead end up in our sewers and have to be handled by wastewater treatment plants. In Europe nearly all wastewater is treated by well-developed plants that can handle most of the wastewater-related problems of the past (infectious diseases, stinking and "dead" rivers due to eutrophication, sewer-related particle emissions, etc.) fairly well. What most of us are not so well aware of is the fact that these plants often do not perform very well at elimination (removal from the water) or even mineralization (i.e., transformation of organic compounds to carbon dioxide, water, etc.) of such manmade compounds. The chemical remains of our civilization still pass these sometimes highly advanced plants, leading to high concentrations of xenobiotics in rivers, which we expect to be clean and often use as a drinking water resource. On the other hand, the water in these rivers contains such high amounts of hormone-like compounds (endocrine-disrupting agents) that considerable numbers of male fish show feminization. In some cases this goes as far as egg production in male gonad tissue [3].

This does not necessarily mean that we live in a world close to catastrophe, but it does imply that our rivers are not as clean as we would like them to be. Additionally, it means that consumers have to pay a high price for the purification of drinking water.

The European Communities have implemented the Water Framework Directive [4], thus showing a strong desire to protect and improve water resources and aquatic ecosystems in the near future. There are indeed several options for improving the chemical quality of surface waters. Further improving wastewater treatment technology, restricting the use of dangerous or risky chemicals to closed systems, and banning groups of chemicals are just some of these options. Making consumers aware of the environmental implications of everyday life's choices may be another.

Drinking water and surface water are big issues in Europe, as is contamination of food, be it via tainted meat from mishandled industrial animal (e.g., pig) production or fruits and vegetables contaminated by pesticides. In the last decade, several incidents have shown that animal feed is contaminated with PCBs, when handling of PCB wastes is not appropriate.

Protective measures currently in use are in principle effective in protecting the consumer from pesticide residues in vegetable and fruit products. However, several modern pesticides as well as steroid hormones and some pharmaceuticals cannot be quantified utilizing the analytical methods that were applied in the past.

This book consists of datasets from diverse projects with different backgrounds. Some are related to wastewater, some to drinking and some to surface and marine waters, while some are pure method development. We hope that an interesting and informative book has been generated that may help fellow scientists, students, and people purely interested in the environmental sciences.

The authors hope with this book to demonstrate that sound science can contribute to determining new problems that may arise as a result of production or lifestyle changes, as well as how these problems can be tackled effectively.

Essen, December 2006   *Kai Bester*

Acknowledgments

This work would not have been possible without friends, support, and cooperation of various kinds.

First of all, I have to acknowledge the Ph.D. students from my group who have done great work:

Michael Schlüsener wrote his thesis on the fate of pharmaceuticals and hormones in the environment and supports the group with MS/MS and computer knowledge.

Jens Andresen, neé Meyer, wrote his thesis on organophosphorus flame-retardants and does a fine job in operating the new GC-MS and bringing some hope to our football team. Both he and Michael Schlüsener co-authored some chapters in this book.

Anke Grundmann wrote a diploma thesis on method development for organophosphates from water samples and thus helped to broaden the abilities of our group beyond TCPP.

Christian Lauer has joined the group for his bachelor thesis on acidic pharmaceuticals. Mirka Jamroszak wrote a great diploma thesis on elimination of xenobiotics in membrane bioreactors. Veronika Bicker just finished a diploma thesis on particulate transports. Several other students are doing their research theses in this group and thereby contribute to the advance of science.

During several phases, this work relied on technical assistance in the laboratory: Gabriele Hardes, Cornelia Stolle, and Jennifer Hardes did excellent jobs as laboratory technicians. They were supported by several students, including Mirka, Christof, Tobias, Monika, Markus, and Martin. A big "thank you" to all of you not only for doing good jobs but also for being such a friendly, cooperative group.

I also acknowledge the assistance in starting the group in Essen that I received from Martin Denecke, as well as his willingness to discuss all the stupid questions that chemists come up with concerning the biology of sewage treatment plants.

I appreciate all the crossroads discussions on all kinds of issues concerning science with the environmental analytical group, especially Roland Diaz-Bone.

The waste management group at the University of Duisburg-Essen supported my joining in very nicely, and I felt very quickly quite at home so thanks to

Personal Care Compounds in the Environment: Pathways, Fate, and Methods for Determination. Kai Bester
Copyright © 2007 WILEY-VCH Verlag GmbH & Co. KGaA, Weinheim
ISBN: 978-3-527-31567-3

Annette, Michaela, Veronika, Jochen, Roland, and Jürgen, as well as Jennifer, Nadine, and Maren. Without the help of Johanna and Maren, many more typos would have survived in the final version of the text.

Anke, Jochen, Jörn, Wilhelm, and Thomas made some tough times much more agreeable.

Though some of the "Hühnerfuss connections" are quite distant in terms of geography, I did and do enjoy the discussions, ideas, and possibilities that this network spreads and generates; thus, special thanks go to Ninja, Stefan, Roland, Robert, Jan, Markus, Sonja, and Heino.

I am also indebted to Prof. Hirner, Prof. Hühnerfuss, and Prof. Widman not only for the opportunity to perform this work but also for their friendly support and their demonstrations (all in their own way) that research groups can be led in a friendly, supportive, relaxed, and still efficient way.

Last, but certainly not least, I have to acknowledge support from the founding agencies, the LUA, and the MUNLV. The discussions with Dr. Stock were especially fruitful, and several very interesting approaches and possibilities stem from this cooperation.

List of Contributors

Jens A. Andresen
Institute of Environmental Analytical
Chemistry
University of Duisburg-Essen
Universitätsstrasse 15
45141 Essen
Germany

Kai Bester
Institute of Environmental Analytical
Chemistry
University of Duisburg-Essen
Universitätsstrasse 15
45141 Essen
Germany

Michael P. Schlüsener
Institute of Environmental Analytical
Chemistry
University of Duisburg-Essen
Universitätsstrasse 15
45141 Essen
Germany

Stefan Weigel
Eurofins Analytik
Research and Development
Neuländer Kamp 1
21079 Hamburg
Germany

Personal Care Compounds in the Environment: Pathways, Fate, and Methods for Determination. Kai Bester
Copyright © 2007 WILEY-VCH Verlag GmbH & Co. KGaA, Weinheim
ISBN: 978-3-527-31567-3

List of Abbreviations

4-MBC	4-methylbenzylidene-camphor
AB	Aeration basin
AHTN	7-Acetyl-1,1,3,4,4,6-hexamethyl-1,2,3,4-tetrahydronaphthalene, e.g., Tonalide®
AHDI	6-Acetyl-1,1,2,3,5-hexamethyldihydroindene, e.g., Phantolide®
amu	Atomic mass unit
ANOVA	Analysis of variance
APCI	Atmospheric pressure chemical ionization (in HPLC-MS)
ATII	5-Acetyl-1,1,2,6-tetrametyl-3-isopropyl-dihydroindene, e.g., Traseolide®
ASE	accelerated solvent extraction
BCF	Bioconcentration factor
BCF_L	Bioconcentration factor referring to lipid concentration
BCR	Bureau Communautaire de Reference
BGB 172	GC phase
BLMP	Bund-Länder Messprogramm (national German monitoring program of the North Sea)
bp	Base peak (highest signal in a mass spectrum)
BT	Benzothiazole
BP-3	Benzophenone-3,2-hydroxy-4-methoxyphenylmethanone, oxybenzone
C	Concentration
CB	Carbendazim (pesticide)
CEFIC	European Chemical Industry Council
CID	Collision-induced dissociation
CLA	Clarithromycin
COD	Chemical oxygen demand
CPS	Counts per second
CRM	Certified reference material
DB-5MS	GC phase
DAD	Diode array detector (for HPLC)
DCM	Dichloromethane

Personal Care Compounds in the Environment: Pathways, Fate, and Methods for Determination. Kai Bester
Copyright © 2007 WILEY-VCH Verlag GmbH & Co. KGaA, Weinheim
ISBN: 978-3-527-31567-3

DDT	Dichlorodiphenyltrichloroethane (IUPAC:2,2-bis-[4-chloroben-zene]-1,1,1-trichloroethane)
DEET	Diethyltoluamide
DES	Diethylstilbene
$D_{27}TnBP$	Perdeuterated tri-n-butyl phosphate
EC	Enantiomeric composition
EC_{50}	Effective concentration for 50% of the tests (organisms)
ECD	Electron capture detection
ED_{50}	Effective dose for 50% of the tests (organisms)
EHMC	(2-Ethyl)hexyl-,4-methoxy cinnamate
EI	Electron impact (ionization in mass spectrometry)
EPA	Environmental protection agency (of the USA)
ER	Enantiomeric ratio
ERY	Erythromycin
EU	European Union
ESI	Electrospray ionization
FST	Final sedimentation tank (of a sewage treatment plant)
FPD	Flame photometric detector (for GC)
GC	Gas chromatography
GPC	Gel permeation chromatography (size-exclusion chromatography)
GREAT-ER	Geography-referenced regional exposure assessment tool for European rivers
HCB	Hexachlorobenzene
HCH	Hexachlorocyclohexane (besides lindane (γ-HCH), the a, β, and δ isomers are abundant in environmental samples
HHCB	1,3,4,6,7,8-Hexahydro-4,6,6,7,8,8-hexamethylcyclopenta-(g)-2-benzopyran, e.g., Galaxolide®
HHCB-lac	HHCB-lactone (galaxolidone)
HPLC	High-performance liquid chromatography
hRT	Hydraulic retention time (e.g., of an aeration basin)
I	Relative intensity (in spectra and chromatograms)
IAL	IAL Consultants, London
IEV	Inhabitant equivalent factor
i.d.	Inner diameter
ID	Isotope dilution
IDMS	Isotope dilution mass spectrometry
IHCP	Institute for Health and Consumer Protection (of the European Commission)
Inchem	Database on chemicals hosted by the World Health Organization, the United Nations Environment Program, and the International Labor Organization (www.inchem.org)
IRMM	Institute for Reference Materials and Measurements (of the European Commission)
IS	Internal standard (for gas chromatographic quantification)
ISO	International Organization for Standardization

IST	Intermediate sedimentation tank (of a sewage treatment plant)
ITD	Ion trap detector (mass spectrometer)
IUCLID	International Uniform Chemical Information Database (of old chemicals)
LAS	Linear alkylbenzene sulfonates
LLE	Liquid–liquid extraction
LOD	Limit of determination
LOQ	Limit of quantification
$M^{\bullet *}$	Molecular radical ion
MCF-7	Human breast cancer cell line
μM	Micromole
Me	Methyl group
MRM	Multi-reaction monitoring (SRM/selected reaction monitoring)
MS	Mass spectrometry
ms	Millisecond
MS/MS	Tandem mass spectrometry
MTB	Methythiobenzothiazole
MX	Metalaxyl
na	Not analyzed
nd	Not determined
NIST	National Institute of Standards and Technology (of the USA)
NP	Nonylphenol
NP-HPLC	Normal-phase HPLC
NPD	Nitrogen/phosphorus-selective detector (for GC)
NPEO	Nonylphenol ethoxylate
OC	Octocrylene
O.D.	Outer diameter
OECD	Organization for Economic Cooperation and Development
OSPAR	Oslo and Paris Commissions to Protect the Atlantic Ocean and the North Sea
OTNE	[1,2,3,4,5,6,7,8-Octahydro-2,3,8,8-tetramethylnaphthalen-2yl] ethan-1-one (= Iso E Super)
OV1701	GC phase
PAH	Polyaromatic hydrocarbons
PBDE	Polybrominated biphenyl ether
PCB	Polychlorinated biphenyl
PCCH	Pentachlorocyclohexene
PFTBA	Perfluorotributylamine
pK_{ow}	Logarithm of the octanol–water partition constant
pM	Picomole
POP	Persistent organic pollutant
PPC	Pharmaceuticals and personal care products
PSI	Pounds per square inch
PST	Primary sedimentation tank (of an sewage treatment plant)
PSU	Practical salinity units

PTFE	Polytetrafluoroethylene
PTV	Programmable temperature vaporizer (GC injector)
PUF	Polyurethane foam
ROX	Roxithromycin
RP-HPLC	Reversed-phase HPLC
rr	Recovery rate
RSD	Relative standard deviation
RT	Retention time (of a compound in a chromatographic system)
RTX 2330	GC phase
RV	Research vessel (ship)
SD	Standard deviation
SDB	Styrene divinylbenzene
SEC	Size-exclusion chromatography (GPC)
SIM	Selected ion monitoring
SPE	Solid-phase extraction
SPM	Suspended particulate matter
SPMD	Semipermeable membrane device
SRM	selected reaction monitoring (= MRM)
SSI	Split/splitless injection
STP	Sewage treatment plant
TB	Thiabendazole (pesticide)
TBEP	Tris(butoxyethyl)phosphate
TDCP	Tris-(dichloro-*iso*-propyl)phosphate
TCEP	Tris-(2-chloro-ethyl)phosphate
TCMTBT	Thiocyanato-methylthiobenzothiazole (biocide)
TCPP	Tris-(2-chloro-*iso*propyl)phosphate
TiBP	Tri-*iso*-butylphosphate
TIC	Total ion current
TIE	Toxicity identification evaluation
TnBP	Tri-*n*-butylphosphate
TOC	Total organic carbon
TPP	Triphenylphosphate
U	Uncertainty
XAD	SPE extraction material (styrene-divinylbenzene)

1
Introduction

1.1
General Considerations

In this study the focus is on the ingredients in personal care products such as polycyclic musk fragrances, household bactericides, and organophosphate flame-retardants and plasticizers, as well as some endocrine-disrupting agents that have been studied as compounds that are entering the aquatic environment mostly via sewage treatment plants (STPs). All of these compounds are used in the range of several thousand tons annually, most of them in applications near wastewater streams such as washing and cleaning. For the flame-retardants, one of the main issues was establishment of a link to surface water contamination, because, technically, usage and wastewater are not obviously connected. For those compounds that were found to be present in surface waters in more industrialized areas such as the Ruhr metropolis, tests were performed to determine whether degradation possibly happened in the respective river or plant. Enantioselective analysis was used in some cases for chiral compounds to identify biodegradation under the assumption that only biodegradation can result in a chiral shift, i.e., an enantiomeric excess from a racemate (see Sections 2.1.3 and 2.5.2) [5].

In this work the word "degradation" will be avoided, but "transformation" will be used if a reaction from one organic compound to another by any means whatsoever is addressed. "Mineralization" will be used when it is assumed that a compound will be transformed to carbon dioxide, water, etc. "Elimination" will be used to demonstrate that the fate of the compound is unknown but the concentration of the parent compound decreases.

For all of the compounds studied, robust methods were established, and the respective standard deviations and limits of detection are given in the respective chapters. In Chapter 3 method development is discussed in more depth, e.g., some of the flaws that may be encountered while quantifying with HPLC-MS/MS.

In several experiments it is hard to discriminate between dilution of xenobiotics in the (aquatic) environment, sorption to particles and sediments, and transformation processes. To discriminate between dilution into open waters and other processes, markers can be used as demonstrated in the sections on marine pollution. In marine ecosystems, salinity is a suitable marker, as most compounds are brought into the sea by freshwater streams.

For risk assessment, diverse pieces of legislation are currently relevant. For limnic systems, the most important one on the European scale is the Water Framework Directive [4]. On the national scale, diverse regulations result in target values regarding surface waters. Target values for the limnic situation mostly combine emissions, concentrations, and persistence, and toxicology data are combined for a risk assessment. By this approach, basic data are obtained to define target values. For the marine ecosystems of the North Sea, the Oslo and Paris Commissions for the protection of the North Sea and the Northern Atlantic, respectively, have defined different regulations. The precautionary principle is often used because gaps in data concerning concentrations, persistence, and fundamental knowledge of the ecosystems are more common than in limnic ecosystems.

All in all, "zero emissions" with regard to the marine environment are requested by the OSPARCOM regulations [6]. The intention of this study is to present data on emissions via sewage treatment processes into the rivers, to demonstrate the persistence of some compounds, and to obtain data on the introduction of some of these compounds into the marine ecosystems. For this approach methods for trace and ultra-trace analysis were developed.

This study was performed to give new insights into elimination mechanisms of xenobiotics in sewage treatment as well as to study the persistence of organic compounds in limnic and marine ecosystems. To study elimination mechanisms of xenobiotics from wastewater, mass balances including sorption of compounds to the sludge were performed. Thus it was possible to discriminate between mineralization/transformation and pure sorption to sludge. Whenever applicable, assumed transformation processes were included in this study for holistic mass balances.

1.2
Introduction to Sewage Treatment Plant Functions

Today's sewage treatment plants (STPs) are designed to eliminate particulate material. The major task, however, is to eliminate organic carbon such as that expressed in the parameter total organic carbon (TOC) or the more biologically defined biological oxygen demand (BOD). The target was thus to prevent the receiving waters from becoming anaerobic. Additionally, most plants have also been equipped with nitrogen and phosphorus removal processes to prevent eutrophication. They have never been designed to control the emissions of priority pollutants or other persistent organic compounds.

TOC removal is realized in most STPs by aerobic activated sludge treatment in which the dissolved organic compounds are transformed into carbon dioxide and biomass. The biomass is then separated and treated in anaerobic sludge treatment before final disposal.

Nitrogen is removed by oxidizing ammonia, which is toxic to fish, to nitrate and either including this into the biomass or reducing it to elemental (gaseous) nitrogen. For this process the medium needs to be anaerobic, which is classically performed in an upstream denitrification. However, in real-life wastewater treatment, simultaneous denitrification is used with aerated and non-aerated bands or areas in activated sludge treatment.

Phosphorus removal is normally performed as precipitation with iron salt solutions. In most cases simultaneous phosphorus elimination is performed, and thus the iron salt solutions are added to the main treatment basin.

During the passage of the wastewater, the water experiences different ecological situations and predominant bacterial communities; thus, it is hard to predict in which part of the sewage treatment which processes might be relevant for a given anthropogenic and possibly unwanted compound.

A schematic sketch of STP functioning is given in Fig. 1.1.

Typical elimination pathways include:

1. Sorption to sludge (biomass). Primary sludge should normally contain higher concentrations than excess sludge.
2. Oxidative transformation, especially in the aerated parts of the activated sludge treatment. Ideally, the final products of this process should be carbon dioxide, etc.
3. Reductive transformation, especially in the non-aerated parts of the activated sludge treatment. This might be especially relevant for dehalogenation processes. However, these processes are normally too slow to be performed within a few hours residence time of the activated sludge treatment.

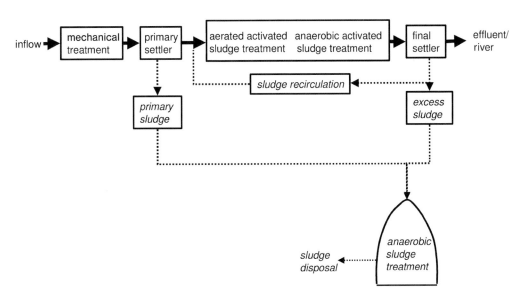

Fig. 1.1 Basic functionalities in a sewage treatment plant.

1.3
Enantioselective Analysis in Environmental Research

Several xenobiotic compounds – including pesticides such as chlordane, toxaphene, and metalaxyl; pesticide impurities such as a-HCH; and synthetic fragrances – are chiral compounds (see Fig. 2.1) [5]. Most of these compounds are supposed to interact with chiral biological receptors because of their desired biological effects. It has been shown that biotransformation reactions of such compounds in vertebrates, as well as in sediment–water systems (as a result of microorganisms), often are enantioselective processes, as a multitude of enzymes that take part in these transformation processes perform the respective reactions with considerable enantioselectivity [7–12] .

Considering instrumental analysis, it should be kept in mind that in some cases apparent racemates do not give peak ratios 1:1 at all concentrations. Enantioselective calibration is thus essential for all enantioselective chromatographic systems.

A very good overview on enantioselective separation systems is given by Ward [13], who also indicated that enantioselective gas chromatography (GC) is still a more dynamic research area in comparison to enantioselective high-performance liquid chromatography (HPLC) systems or other separation methods. A good overview on enantioselective analysis for environmental issues is given in Ref. [8].

1.3.1
Enantioselective Gas Chromatography Techniques

Enantioselective GC phases are based mostly on cyclodextrins, which have become more and more available and stable in the last few years. Long retention times of about 30–90 min still need to be taken into account, as the solvation enthalpy differences of the respective enantiomers often are small. These long retention times thus have relatively broad peaks (about 20 s), while in conventional capillary GC, peak widths of 4 s are experienced. These broader peaks consequently give lower detection limits. On the other hand, the precision of the determination is limited only to chromatographic overlaps and the precision of the integrating system. This is due to the fact that the enantiomers behave physically absolutely identical. This means that recovery rates, evaporation, and sorptive losses, etc., are identical for both enantiomers, as long as no chiral materials are used in the sample preparation scheme. Standard deviations of 1% and less are regularly obtained with established separation systems.

No chemically bonded phase is commercially available nowadays. Thus, temperature stability of the GC phase is gained by mixing the enantioselective discriminator, such as the cyclodextrin derivative, with other phases, such as OV-1701. These phases can operate at limits of up to 230 °C, which is an enormous improvement considering the situation five years ago. This limit still inhibits classical baking of the enantioselective columns in comparison to, e.g., DB-5 columns, which exhibit temperature limits of up to 400 °C. Thus, for enantioselective analysis more-selective cleanup procedures than in conventional analysis are needed. Enantioselective GC equipped with mass spectrometric (MS) [9] or electron capture detection (ECD) [7] has been used to determine enantioselective degradation of organochlorine pesticides in vertebrates, thus giving good evidence for biodegradation or biotransformation of such compounds in, e.g., marine mammals. Typically, two-step cleanups, e.g., consisting of size-exclusion and silica-sorption chromatography, are used for enantioselective determinations. Especially for ECD analysis, additional normal-phase HPLC fractioning for sample preparation was necessary in some cases [14]. Fractionation is especially important for the enantioselective analysis of toxaphenes, as the original toxaphene pattern is extremely complex. Thus, enantioselective separation has to be performed in combination with a classical separation of some hundred congeners [14]. It must be taken into consideration that there is no such thing as "the enantioselective GC phase." Some columns separate several compounds easily but fail on very similar substances. An overview on separations of relevant chiral pollutants that have already been separated is given in Table 1.1.

Table 1.1 GC phases for separation environmentally relevant enantiomers.

GC phase	Trade name	Analytes separated
Heptakis(3-O-butyryl-2,6-di-O-pentyl)-β-cyclodextrine	–	a-HCH, PCCHs [8]
Heptakis(2,3,6-tri-O-n-pentyl)-β-cyclodextrine in 50% OV1701	Lipodex C	a-HCH, β-PCCH [8]
Heptakis(2-O-methyl-3,6-dipentyl)-β-cyclodextrine	–	Oxychlordane, *cis*-heptachlor epoxide [8
Heptakis(6-O-*tert*-butyldimethylsilyl-2,3-di-O-methyl)-β-cyclodextrinin in 20–50% OV1701	Hydrodex; BGB172	Bromocyclene [10], PCB 88, PCB149, PCB183, PCB171, PCB 174 [8], oxychlordane, *trans*-heptachlor epoxide, allethrin, bioallethrin, methamidophos, acephate, trichlofon, bromacil, PCB 45, 95, 91, 136, 131, 176, 175 [16] dimethenamid, metalaxyl, metolachlor [11] HHCB, AHTN [35] ATII, AHDI [12], o,p'-DDT [17], methylated mecoprop [18]
Octakis(3-O-butyryl-2,6-di-O-pentyl)-γ-cyclodextrine in 50% OV1701	Lipodex E	a-HCH, *trans*-chlordane, PCB95, PCB136 [19]
Octakis(2,6-methyl-3-pentyl)-γ-cyclodextrin (in 80% OV1701)	–	a-HCH, *cis*-chlordane, *trans*-chlordane [16]
Heptakis(2,6-methyl-3-pentyl)-β-cyclodextrin (in 80% OV1701)	–	Heptachlor, *cis*-heptachlor epoxide [16]
Heptakis(2,3,6-trimethyl)-β-cyclodextrin with some *tert*-butyldimethyl substituents	–	Toxaphenes [9, 14]
Heptakis(2,3,6-tri-O-*tert*-butyldimethylsilyl)-β-cyclodextrin coupled to RTX 2330	–	*cis*- and *trans*-chlordane, *trans*-nonachlor [20]

1.3.1.1 Applications of Enantioselective Gas Chromatography

Determination of Biodegradation
Enantioselective GC has been used extensively to determine whether or not bio-degradation is relevant in selective media such as vertebrate tissue, surface water, sediment, tissue, sewage sludge, soil, etc. [7–12, 14, 15]. These experiments work well under the assumption that only enzymes perform enantioselective reactions in the environment. Thus, if an enantiomeric excess is determined, a biodegradation is highly probable.

Determination of Phase Transfer of Pollutants
For some time it was assumed that, e.g., HCHs might evaporate from the Great Lakes in the U.S. and Canada. It is extremely difficult to prove this assumption based on Henry's law, and it is difficult to analyze these compounds at levels of nanograms per liter in the water or picograms per cubic meter in the air. A mass transfer for such huge ecosystems is thus very hard to determine. On the other hand, knowledge of such processes is essential for the assessment of transport of these organochlorine compounds into the Arctic. Ridal et al. [21] found that a-HCH exhibited a peculiar enantiomeric distribution in the water of Lake Ontario, which could be determined with extreme precision: about 1% standard deviation was found for the determination of enantiomeric ratios as sample-to-sample deviation, as well in air, rain, and surface water samples. In comparison, rainwater samples, which were taken as a measure of the enantiomeric ratio of a-HCH in the higher atmospheric layers because the droplets were formed in high altitudes, were found to contain racemic a-HCH. Additionally, the enantiomeric ratio of this compound was measured in air samples taken from sea level as well as from lake water. Because the higher levels of the atmosphere (rain) contained true racemic composition, and the enantiomeric ratios of a-HCH in the air samples in summertime were very similar to the water, it could be concluded that indeed in summertime a vaporization of a-HCH from the water occurred. However, in wintertime the situation may be different.

Determining the Dominant Sources of Pollution
Mecoprop is a chiral phenoxyalkanoic acid herbicide that is marketed as an enantiopure compound for agriculture (pesticide). Its levels in Swiss surface waters are moderate but are surprisingly high considering that it has only agricultural applications. In 1998 it was found that the same compound was used in roof materials to prevent plants from growing on top of flat roofs. In contrast to the agricultural applications, the mecoprop used for roof sealing is marketed as racemate. Thus, the surface water samples were analyzed for the enantiomeric ratios. Because the enantiomeric ratio was about 0.5 in environmental samples, while 0 for agriculture and 1 for the rooftop material, it was possible to determine that about 50% of the mecoprop in Swiss surface water originated from rooftops and not from agriculture [18].

1.3.1.2 **New Developments**

Currently enantioselective GC is used, e.g., to determine whether or not chiral synthetic fragrances such as polycyclic musk fragrances are possibly biodegraded or whether adsorptive processes dominate the elimination in sewage treatment plants. These compounds bioaccumulate in fish; therefore, higher elimination rates in the respective wastewater treatment processes are urgently sought after. Chirality could give an indication as to which parameter (aerated biologically activated sludge, anaerobic treatment, or sorption phenomena) in the plant should be optimized.

In this study enantioselective analysis was performed as gas chromatographic separations for the synthetic fragrances HHCB, AHTN, and HHCB-lactone, as well as for the insecticide bromocyclene to observe transformation processes in sewage treatment plants (see Sections 2.1.3 and 2.5.2).

1.3.2
Enantioselective HPLC

Enantioselective HPLC is also used in environmental studies [7], though the major applications of enantioselective HPLC separations at the moment are in the field of drug development. This is probably due to the fact that the separation power of enantioselective HPLC columns with regard to the complex environmental matrix is somewhat limited. Therefore, the risk arises that the true enantioselective separation overlaps with compounds in the matrix, thus giving unreliable results. High selectivity of the respective detectors, such as tandem mass spectrometry, and a well-known matrix, as in a controlled soil degradation experiment, are thus essential prerequisites for the application of enantioselective HPLC columns in environmental sciences. It should also be noted that a multitude of different separation mechanisms are currently utilized in HPLC. In any case, the possibilities of combining columns with eluents are fascinating.

1.3.2.1 **Applications of Enantioselective HPLC**

Metalaxyl, metolachlor, and alachlor are chiral pesticides that have been marketed as racemates, while only one stereoisomer gives most of the desired biological effects (herbicides and fungicides). Some of these compounds cannot readily be separated by enantioselective GC but can easily be separated by HPLC, e.g., on a Whelk-O 1 column. Nowadays these studies are used to estimate the fate of both enantiomers in diverse ecosystems, especially in soil, to determine whether or not there are differences under diverse climatic and ecological situations [11, 15].

Enantioselective analysis used to be a method that could be used only by very specialized laboratories for fancy purposes. This situation has changed in the last few years to a method that any laboratory that has some experience in chromatography and sample pretreatment can use with reasonable effort. New insights into biodegradation as well as transport phenomena can be gained from this technique. Enantioselective analysis is thus a dynamic field bridging issues from environmental sciences, bio and life sciences, metabolomics, and analytical chemistry.

2
Environmental Studies: Sources and Pathways

2.1
Synthetic Fragrance Compounds in the Environment

Polycyclic musk compounds such as HHCB (1,3,4,6,7,8-hexahydro-4,6,6,7,8,8-hexamethylcyclopenta-(g)-2-benzopyran; trade name, e.g., Galaxolide®) and AHTN (7-acetyl-1,1,3,4,4,6-hexamethyl-1,2,3,4-tetrahydronaphthalene, trade name, e.g., Tonalide®) are used frequently as fragrances in washing powders, shampoos, and other consumer products that are supposed to smell pleasantly. More than 2000 t are used annually in Europe [22]. The structural formulas of both compounds as well as the transformation product HHCB-lactone are given in Fig. 2.1. Most of the compounds are eventually disposed of via the wastewater stream. Thus, they have been identified in sewage treatment plants in Europe and the U.S. and in freshwater in Europe [23–27, 38]. An overview on musk fragrances in the environment is given by Rimkus [28], addressing mostly human exposure. However, neither total balance on STP processes, nor mass transport data in rivers, nor marine data were included in those studies. Additionally, polycyclic musks have been determined in the waters of the North Sea [29] (see also Section 2.1.4). These compounds have also been analyzed in a variety of freshwater fish in Europe and Canada by Gatermann et al. [30, 31], as well as in human tissue by Rimkus and Wolf [32]. The concentrations in surface waters ranged around 100 ng L^{-1} at that time. These polycyclic musk compounds exhibit high bioaccumulation power (log K_{ow} ~6), and thus high concentrations in fish are easily explained. A study in 1999 by Seinen et al. [33] showed that HHCB and AHTN exhibit some estrogenic effects, which gave even more reason for concern. Simonich et al. [25] published a study on fragrances in sewage treatment plants in which they found removal of 90% for an activated sludge treatment and 83% for a trickling filter plant in Ohio for both compounds; no data on sludge were presented in that study. In another study, the same group [26] studied a multitude of sewage treatment plants and found elimination rates ranging from ~50% to 95%. Most plants with high elimination efficiency were rather small ones (20 000 m^3 d^{-1} and 1500 m^3 d^{-1}) that operated with domestic wastewater only. In more densely populated Germany, a plant operating at ~200 000 m^3 d^{-1} is considered medium sized. Thus, it was decided to study the fate of HHCB and AHTN in one of the larger German plants. To obtain a full balance, sludge was also ana-

Personal Care Compounds in the Environment: Pathways, Fate, and Methods for Determination. Kai Bester
Copyright © 2007 WILEY-VCH Verlag GmbH & Co. KGaA, Weinheim
ISBN: 978-3-527-31567-3

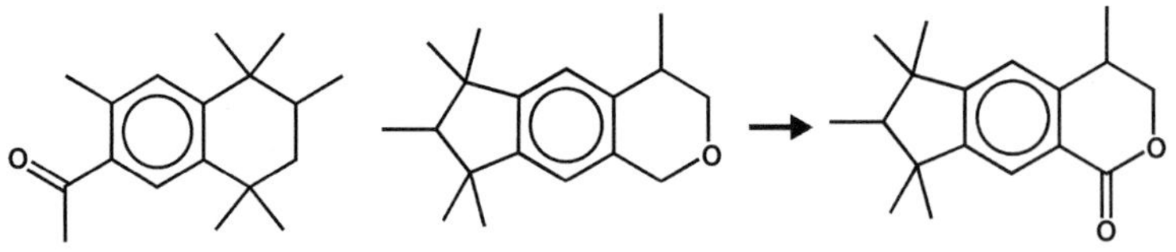

AHTN (Tonalide)   HHCB (Galaxolide)   HHCB-lactone
                                                   (galaxolidone)

Fig. 2.1 Structural formulas of AHTN, HHCB, and the primary metabolite HHCB-lactone.

lyzed to determine whether the observed elimination was due to sorption or bio-transformation. Sludge was not studied in the work of Simonich et al. [25, 26]. In the presented study, the primary metabolite of HHCB, which is an oxidation product, i.e., HHCB-lactone (see Fig. 2.1), was also included. This metabolite was identified in fish samples by Kallenborn et al. [34] and by Franke et al. [35] in water samples. The question was thus whether this compound originated from metabolic processes in fish or river sediments or from the sewage treatment processes. This study presented here was performed to find elimination rates for a large, mixed-purpose plant as well as to determine whether sorption or degradation was the dominant process for the removal of the polycyclic musk fragrances.

2.1.1
Polycyclic Musk Fragrances in Sewage Treatment Plants

2.1.1.1 Experimental Background
Because polycyclic musk compounds are used in washing powders, shampoos, etc., they reach the STPs shortly after application. It is currently a point of discussion to what extent these compounds are eliminated with current STP technology. It is unclear by which mechanism an elimination process may take place. Thus, in this part of the study the fate of AHTN and HHCB in an STP was monitored not only for in- and outflow of water but also for the fraction of these compounds sorbed to the sludge that is exported from STPs to incinerators or used as fertilizer on agricultural land (fields) [68].

Samples were taken at an STP located in the vicinity of Dortmund, Germany, that processes $200\,000$ m^3 wastewater per day. This plant processes the wastewater of about $350\,000$ inhabitants as well as of industry, mainly breweries. About half of the wastewater that is processed is domestic. Thus, the water of the brewery, which is not supposed to contain relevant amounts of polycyclic musk compounds, potentially dilutes the water contaminated with polycyclic musk fragrances from domestic sources. Sewage treatment plants operating on a mix of domestic and industrial wastewater as well as the size are typical for the Ruhr megalopolis consisting of 5–7 million inhabitants. The plant is an activated sludge plant with secondary treatment. It includes primary settlement basins, activated sludge treatment basins (aeration basins), simultaneous nitrogen removal, sludge separation basins, anaerobic digesters, and a final clarifier before the water is released to the river. After anaerobic digestion the sludge is treated with a filter press to dewater the sludge. Therefore, the final material contains about 70% water. The water from the filter press is brought back into the aeration basin. The water temperature in the influent was 10 °C, while that in the effluent was 14 °C. Total suspended solids were recorded in volumes in Germany by a sedimentation experiment. About 9 mL L^{-1} in the influent and below 0.1 mL L^{-1} in the final effluent were determined within this experiment. Chemical oxygen demand (COD) in the inflowing water was 410 mg L^{-1}, with about 11% day-to-day RSD. The plant operated in steady state without rainfall for several weeks before the sampling started. Hydraulic retention in the aeration basin was 8 h, while sludge retention was 8–10 d. Sludge retention in the digester was 20 d at 37 °C.

The experiment was performed from 8–12 April 2002. Water samples (1 L) were taken every two hours automatically and mixed to give 24-hour composite samples, which are relevant for the local authorities. Thus, continuous time-proportional sampling was performed for this 120-h experimental period. Each day, two of these samples for inflowing and outflowing water were taken as duplicates. Inflow samples were taken after the water had passed the mechanical particle separation (grid chamber). Outflowing water was sampled after the water had left the final settlement basins before introduction into the receiving water, i.e., the respective river. The samples were stored at 4 °C during the sampling and were extracted within 4 h after the finalization of the 24-h sampling cycle. Thus, no preservative was necessary. The whole procedure is described in detail in Section 3.1. In brief, the samples were extracted with toluene, condensed, and analyzed by means of GC-MS.

This procedure was validated with pure water and gave recoveries of 75–78%. Full quality data of the method obtained from three replica extractions at five different concentrations are given in Table 2.1. The limit of quantification (LOQ) was established by the recovery experiments as the lowest concentration at which the recovery rate was within the described range (i.e., 78±7% for AHTN, 75±6% for HHCB, and 100±23% for HHCB-lactone). Thus, the working range was LOQ to 10000 ng L^{-1} (Table 2.1). The LOQ was 10 ng L^{-1}, 100 ng L^{-1}, and 5 ng L^{-1} for

Table 2.1 Method quality data for the extraction of musk compounds from aqueous samples.

	Quantifier mass (amu)	Verifier mass (amu)	RT (min)	rr (%)	SD (%)	RSD (%)	LOQ (ng L^{-1})
AHTN	243	258	9.52	78	7	9	10
HHCB	243	258	9.39	75	6	8	100
HHCB-lactone	257	272	14.80	100	23	23	5

RT: retention time; rr: recovery rate; SD: standard deviation; RSD: relative standard deviation; LOQ: limit of quantification; amu: atomic mass units.

AHTN, HHCB, and HHCB-lactone, respectively. The blank value (procedural blank) for water samples was determined to be 30 ng L^{-1} for HHCB and 3 ng L^{-1} for AHTN. The more commonly used criterion for the limit of quantification, i.e., 10 times the standard deviation of the blank, would result in LOQs that are slightly lower than the ones used in this paper. However determined, the LOQ is much lower than the lowest concentration determined in this study.

The method was also tested for influence of changes in pH (5–9), humic compounds, and detergent concentrations. The recovery rate did not change under these conditions. In Fig. 2.2 a comparison of retention times of standards and a wastewater inflow sample is shown.

Sludge samples from the anaerobic digester were taken as single-grab samples on the same days as the water samples; they were obtained from the loading of trucks. They were thus the final solid product after anaerobic digestion and dewatering by filter press, containing about 60% water. The samples were immediately stored at 4 °C until extraction, which was performed the same week as the sampling. The whole procedure is described in detail in Section 3.3. Briefly, 10-g samples were Soxhlet-extracted with ethyl acetate, and the extracts were condensed and cleaned up by a combination of size-exclusion and silica-sorption chromatography.

Recovery rates (rr), limit of quantification (LOQ), and standard deviations (SD) were obtained by 15 extractions with spike concentrations ranging from about 5 ng g^{-1} to 1000 ng g^{-1} for each compound. The limit of quantification was established as the lowest concentration in accordance with the recovery rates obtained from the higher concentrations (Table 2.2). Blank (procedural blank) concentrations for both AHTN and HHCB were 2 ng g^{-1}.

All standard compounds were obtained from Ehrenstorfer (Augsburg, Germany), AHTN as a pure compound and HHCB as technical grade (50% purity).

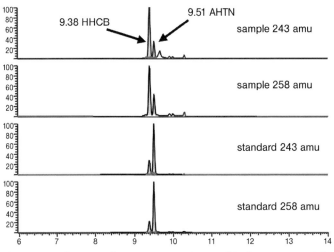

Fig. 2.2 Mass fragment chromatogram (in SIM mode) of a wastewater inflow sample extract.

Table 2.2 Recovery rates (rr) from model sludge of manure-soil mixture.
Also given are standard deviation (SD), relative standard deviation (RSD),
and limit of quantification (LOQ).

	rr (%)	SD (%)	RSD	LOQ (ng g^{-1})
AHTN	76	13	0.17	6
HHCB	100	28	0.28	5

All values are corrected for this impurity. The internal standard D$_{15}$ musk xylene was obtained as a solution from Ehrenstorfer. HHCB-lactone was received as a pure standard as a gift from International Flavours and Fragrances (IFF) (Hilversum, the Netherlands).

Ethyl acetate and cyclohexane were analytical-grade (p.a.) quality, while toluene and *n*-hexane were residue-grade (z.R.) quality. All solvents were purchased from Merck (Darmstadt, Germany).

2.1.1.2 **Mass Balance Assessment**

In the sewage treatment plant under observation, the inflow concentrations of HHCB were about 1900 ng L^{-1}, while those for AHTN were about 580 ng L^{-1} (see Table 2.3). In this table, double samples (24-h composite) for both inflow and effluent were compared with double extractions of sludge samples. These inflow and effluent data are in good agreement with those published by Eschke [23, 24], who analyzed inflow of wastewater in STPs. In that study about 1500 ng L^{-1} were determined as an average from 20 STPs. Concentrations similar to those in the current study and in earlier German studies were determined in Canadian and Swedish wastewater samples [36]. However, Simonich [25] found higher concentrations of 10000 ng L^{-1} in the U.S. This may be due to the fact that in Simonich's study the STPs operated on domestic water only: no per-capita emissions were calculated in these U.S. studies. (In Dortmund, the domestic wastewater is "diluted" by brewery wastewater.) Another possibility is that more personal care products are used in the U.S. than in Germany. How-

Table 2.3 Concentrations of AHTN, HHCB, and HHCB-lactone in inflow
and effluent of a German STP (average of 5 days). Additionally, the day-
to-day variation is given as standard deviation. Data were derived from
duplicate samples of 5 successive days each.

	Influent (ng L^{-1})	Effluent (ng L^{-1})	Elimination rate (%)	Breakthrough (%)
AHTN (Tonalide)	580 (±100)	210 (±17)	63 (±10)	37 (±5.9)
HHCB (Galaxolide)	1900 (±350)	700 (±58)	63 (±11)	37 (±6.6)
HHCB-lactone	230 (±40)	370 (±34)		162 (±22)

ever, the concentrations are in the same range as described by Heberer [37] for Berlin, Germany. Similar results in precision and day-to-day variation were obtained for triclosan from the same samples for the same plant [38] (compare Section 2.4).

The day-to-day variation from the data obtained from the Dortmund plant is calculated as a standard deviation from duplicate samples of five successive days, i.e., 120 h, and is also given in this table. Generally, this variation was less than 20%. However, the concentration in the plant's effluent was significantly lower. This resulted in elimination rates of about 60% for both compounds from the water. In comparison to the AHTN/HHCB pattern in earlier years, which was about 1:1, more HHCB is found in the samples, indicating some shift in application pattern [39]. Interestingly, the HHCB-lactone concentration increased from about 230 ng L^{-1} (inflow) to 370 ng L^{-1} (effluent). This increase is highly significant.

Thus, some of the HHCB is obviously transformed (oxidized) to HHCB-lactone during the sewage treatment process. Some HHCB-lactone is contained in the technical Galaxolide product, as it can be found at the entrance of the plant, as well as in some batches of the raw product that were obtained from Ehrenstorfer. Possibly, further transformation may happen in the sewer system. However, the pattern in the STP's inflow is similar to the one found in the technical product. A full EI mass spectrum of HHCB-lactone obtained from a wastewater sample extract is given in Fig. 2.3.

During the sampling period of water and sludge samples in parallel, the total flow of water was also monitored. About 184 000 m^3 water was flowing through

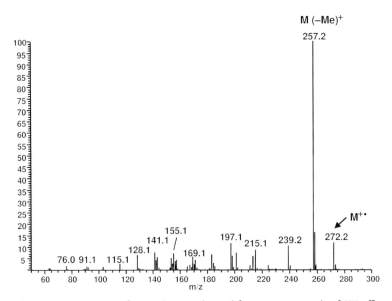

Fig. 2.3 Mass spectrum of HHCB-lactone obtained from a water sample of STP effluent.

the plant each day, with a day-to-day variation of 4100 m^3. Because a predominantly dry season was chosen, the variation was very low.

The concentrations of AHTN and HHCB were also measured in digested, dewatered sludge. For AHTN a medium concentration of 1500 ng g^{-1} dry weight was measured, with a day-to-day standard variation of ± 150 ng g^{-1}, while the concentration of HHCB was 3100 ± 240 ng g^{-1}. These concentrations are at the low end in comparison with other plants in this area, as comparison studies have revealed (see Section 2.1.2). The concentrations showed a very uniform distribution, which made balancing easier. These concentrations are on the low end but are within the same order of magnitude as those cited by Balk and Ford [40]. Similar concentrations have been determined by Herren and Berset [41] (Switzerland) and Lee et al. [42] (Canada). The concentrations in this study are smaller but are within the same order of magnitude as those determined by Kupper et al. [43] for Swiss samples (2004) and those reported by Heberer [37] for samples from Berlin, Germany. To perform a balancing approach, the concentrations in the sludge need to be compared with the amount of sludge that the plant disposed of in the respective time. The management of the plant gave access to these numbers during the respective sampling interval. About 140 t, with a variation of 38 t (27%), sludge was transported from the plant each day. The discontinuous production of dewatered sludge thus gives the largest contribution to the uncertainties for the balance calculation.

A balance was calculated to estimate the pathways of the respective compounds for this five-day period. The respective data are shown in Table 2.4. These balances were calculated on a daily basis and summed up for the five-day period. All values are rounded at 10%. In the final balance, negative contributions indicate losses resulting from either chemical conversion of the respective compound or unaccounted losses, e.g., to the atmosphere. On the other hand, positive contributions indicate sources of the respective compounds in the STP itself that might occur from direct wastewater introduction to the aeration basin, e.g., from wastes of chemical toilets, or from wastes added to the digester

Table 2.4 Balance of HHCB and AHTN in a 5-day sampling period in a German STP. Positive/negative values are day-to-day variations calculated as standard deviations.

Compound	Influent	Effluent	Sludge	Balance	Balance (± SD) (Range)
AHTN (Tonalide)	540 g (100%)	200 g (37%)	430 g (80%)	+87 g (+16%)	(± 28) (−23 to 57)
HHCB (Galaxolide)	1800 g (100%)	640 g (36%)	860 g (48%)	−290 g (−16%)	(± 18) (−40 to 7.4)
HHCB-lactone	210 g (100%)	340 g (162%)	nd	+130 g (>+62%)	

nd: not determined.

Table 2.5 Concentrations of AHTN, HHCB, and HHCB-lactone in influent and effluent as well as in sludge in a German STP (average of 5 days). Additionally, the day-to-day variation is given as standard deviation. The data were derived from duplicate samples.

Date	AHTN				HHCB				HHCB-lactone		
	Influent (ng L^{-1})	Effluent (ng L^{-1})	Break-through (%)	Sludge (ng L^{-1})	Influent (ng L^{-1})	Effluent (ng L^{-1})	Break-through (%)	Sludge (ng L^{-1})	Influent (ng L^{-1})	Effluent (ng L^{-1})	Break-through (%)
08.04.2002	617	240	39	1480	2182	795	36	1480	270	420	156
09.04.2002	713	215	30	1532	2325	691	30	1532	270	370	137
10.04.2002	587	206	35	1343	1933	652	34	1343	230	370	161
11.04.2002	572	203	35	1746	1857	669	36	1746	215	340	158
12.04.2002	427	197	46	1525	1409	669	48	1525	170	335	197
Mean	583	212	37	1525	1941	695	37	1525	231	367	162
SD	103	16.7	5.9	145	352	58	7	145	42	34	22

for co-fermentation. Neither of these pathways was relevant in this STP. AHTN shows an insignificantly positive balance, indicating a simple distribution between sludge and water. Most of the AHTN is transferred to the sludge; thus, a simple sorption mechanism is taking place. The same holds basically true for HHCB. About half of the inflowing material is sorbed to the sludge. Both compounds exhibit a high log K_{ow} of about 5.7–5.9 [44]; thus, the sorption processes are in good agreement with older data. This means that the receiving water, i.e., the river, receives about 37% of the HHCB applied in all usages whatsoever. In the STP in total, a negative balance is detected for HHCB (–290 g) during the 5-day period. This compares to a generation of about 130 g of HHCB-lactone with a day-to-day variation of $5 \, g \, d^{-1}$ (Tables 2.4 and 2.5). Considering these data, it can be assumed that about 7% of HHCB is transformed in the STP to HHCB-lactone.

Simonich et al. [25] published data on the elimination of HHCB and AHTN in an STP in the U.S. The concentrations in the wastewater were significantly higher than those in our study. This difference probably originates from the fact that these plants were described as operating on >90% domestic wastewater. No information on sludge data or metabolites is given in this paper. The lower concentrations in German plants have been discussed in this paper already. In another study [26] elimination rates of fragrance compounds in 17 different plants were compared. Removal rates of 50% to >90% were determined for AHTN and HHCB. The highest removal rates were found in sewage lagoons. The average size was rather small (1000–20 000 m^3, plus two operating at 100 000 m^3) in comparison to the plant studied in Dortmund. All plants had less than 20% industrial wastewater input. Because no data on sludge and metabolites were presented in those papers, the elimination mechanism was not discussed. The removal rates in this study are well within the range published in the report by Simonich [26] for larger plants. Additionally, the new study presented here demonstrates that the main process is sorption to sludge and that only a small percentage of HHCB is transformed into HHCB-lactone. No data pointing at mineralization processes were determined.

Biotransformation therefore is not a dominant mechanism for removing HHCB from wastewater in the STP under observation, as ∼50 (±15)% (HHCB) or ∼80 (±27)% (AHTN) of the inflowing material is sorbed to the sludge and disposed of accordingly. These findings agree well with the sediment–water partition for HHCB/AHTN observed by Dsikowski et al. [45]. About 40% of the inflowing material is released to the surface water. The variability and uncertainty in this study stem from the discontinuous dewatering of sludge. Biodegradation may account to about 5–10% of the total of HHCB. The fact that some plants are able to biodegrade polycyclic musk compounds, as well as the varying elimination rates, shows that optimization of sewage treatment processes may be feasible to lower the discharge of these compounds as well as organic micro-contaminants in general. Ozonization may be a more powerful, but also more costly, tool for removing compounds such as polycyclic musks [46].

Because diverse elimination rates have been detected, it seems important to improve the understanding of elimination of polycyclic musk fragrances and related compounds in STPs over long time periods. If the current sewage treatment process with activated sludge cannot be performed in such a way that degradation plays a major role in the balance of these compounds, the only option for decreasing the emissions of HHCB, AHTN, and HHCB-lactone will be to optimize the sludge:water ratios in STPs in such a way that optimal sorption can take place.

2.1.1.3 Multi-step Process Study on Polycyclic Musks

To verify the results obtained in the STP mentioned above and to obtain more information on what part of the STP process is efficient for elimination and transformation of HHCB and AHTN, a more detailed study was performed at two different STPs. Two were located in the Rhine-Ruhr area (STP A and B) and one was in the eastern part of the Ruhr megalopolis (STP C).

The basic functionalities of both STPs are shown in Fig. 2.4, while those of STP C have already been described in Section 2.1.1.

STP A

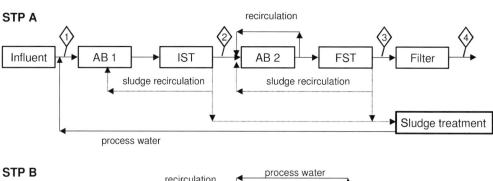

STP B

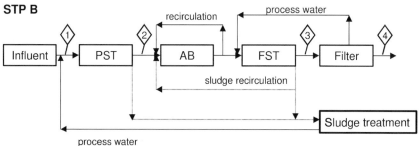

AB: aeration basin
IST: intermediate settlement tank
FST: final settlement tank
PST: primary settlement tank

Fig. 2.4 Basic functionalities of STPs A and B.

Sewage treatment plant A is equipped with a two-stage biological treatment, i.e., two aeration basins with a downstream biological filtration unit (compare Fig. 2.4). Sample site 1 is located at the main collector prior to the sand trap and the screening plant. The process water from sludge dewatering is added before the sampling site. The first aeration basin for the, in this case, highly charged raw wastewater is followed by an intermediate settling tank (IST) before the partially purified water enters the second aeration basin with preceding denitrification. At sample site 2, samples can be gathered from the effluent of the IST. The second biological cleaning step (aeration basin) is followed by the final sedimentation tank (FST). At sample site 3, samples from the effluent of the FST can be received. Before the treated wastewater is discharged to the receiving water, i.e., the Rhine River, it is filtered through a biological filter. The filter bed consists of gravel at the bottom and sand at the top. The treated wastewater and process air for the aeration of the filter flow concurrently from the bottom in an upward direction. Sample site 4 is located at the effluent of the STP after the final filtration unit. Sample sites 2 and 3 were chosen to receive information about the elimination of the analyzed compounds at different stages of the wastewater treatment process, whereas sample sites 1 and 4 provide data on the elimination efficiency of xenobiotics from wastewater.

Sewage treatment plant B is in principle very similar to STP A, but it is a single-stage activated sludge plant with downstream contact filtration. The wastewater flows into the aeration basin with simultaneous denitrification before it enters the primary settling tank (PST). Samples were taken from the influent right after the screening plant and the sand trap (sample site 1) and the effluent of the PST (sample site 2). After the biological purification step, the wastewater is separated from the sludge in the FST. Sample site 3 is located at the effluent of the FST. Finally, the wastewater passes through the contact filtration unit before it is fed into the receiving water, i.e., the Rhine River. The final filter unit is constructed like the one of STP A. Sample site 4 is located at the effluent of the final filtration. Sample site 2 was sampled to study the effect of the primary sedimentation step, whereas sample site 3, in comparison to sample site 4, is supposed to obtain data on the efficiency of the aeration basin and contact filtration in removing xenobiotics from wastewater. Both kinds of wastewater, industrial and municipal, flow into this STP.

Both STP A and STP B are rather large, with wastewater volumes of $108\,959 \text{ m}^3 \text{ d}^{-1}$ at STP B and $220\,000 \text{ m}^3 \text{ d}^{-1}$ at STP A. The corresponding inhabitant values are $1\,090\,000$ for STP B and $1\,100\,000$ for STP A. The main difference results from differing industrial wastewater inflow.

The concentrations of HHCB, AHTN, and HHCB-lactone in the influent of STP A are twice as high as those measured in Dortmund (Fig. 2.5). This is slightly surprising, as both STPs operate at 50% industrial influent. The concentrations decline sharply in the first part (aeration basin 1 + IST) of STP A, while no huge changes in concentrations were observed afterwards for the parent compounds. To obtain more information from these data, actual elimination rates were calculated (Fig. 2.6).

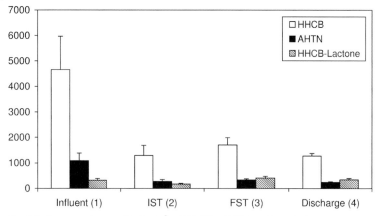

Fig. 2.5 Concentration profile (ng L^{-1}) of HHCB, AHTN, and HHCB-lactone in STP A in influent, after the IST, after the FST, and in the discharge. Error bars indicate day-to-day variations.

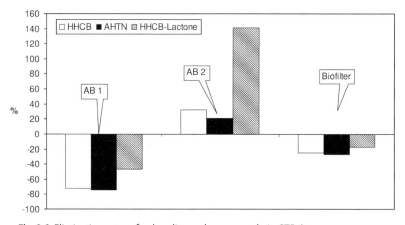

Fig. 2.6 Elimination rates of polycyclic musk compounds in STP A. Negative values indicate elimination processes; positive values indicate generation processes. AB: aeration basin.

The most efficient part of STP A for removal of HHCB and AHTN was the primary aeration basin together with the intermediate settlement tank, in which the concentration profile (average from several days) is shown. About 70% of the initial HHCB and AHTN was eliminated in this step, while about 50% of the HHCB-lactone that entered the STP as an impurity of the product was eliminated in this primary step. The main aeration process (second aeration basin) with preceding denitrification did not have a significant effect on HHCB and AHTN (considering the variabilities determined in this STP; Fig. 2.5), while significant amounts of HHCB-lactone were generated from HHCB in this step. At this point about 20% of the HHCB was transformed to lactone. Finally, the

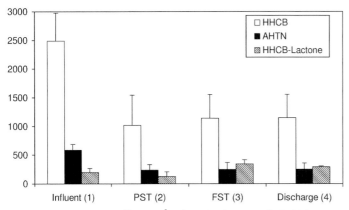

Fig. 2.7 Concentration profile (ng L^{-1}) of HHCB, AHTN, and HHCB-lactone in STP B in influent, after the PST, after the FST, and in the discharge. Error bars indicate day-to-day variations.

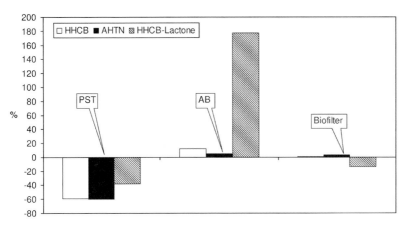

Fig. 2.8 Elimination rates of polycyclic musk compounds in STP B. Negative values indicate elimination processes; positive values indicate generation processes. PST: primary settlement tank, AB: aeration basin.

last filtration step did not change the situation drastically, although slight amounts of musk compounds may have been eliminated.

The *concentrations* determined in STP B were similar to those in Dortmund, but the concentration *pattern* was similar to the one determined in STP A, though this plant was missing the primary aeration basin. A considerable decrease in concentrations was experienced in the primary settlement tank, while no further elimination of the parent compounds was determined (Fig. 2.7) at later stages.

To obtain more detailed data on the respective processes, the elimination rates were calculated (Fig. 2.8). About 60% elimination of HHCB and AHTN was determined, while the amount of HHCB-lactone was reduced with 40% efficiency in the primary step. No further reduction of HHCB and AHTN was determined at any further step of the process. On the other hand, the concentrations of HHCB-lactone nearly tripled, from 120 ng L^{-1} to 344 ng L^{-1}, in the main aeration basin, i.e., about 20% of the HHCB present at this step was transformed. Because of high day-to-day variations, this transformation from HHCB to HHCB-lactone was not reflected by a significant decrease in HHCB concentrations. There was a trend for declining metabolite concentrations caused by the final filter, but, because of day-to-day variability, this was not significant.

It is interesting to note that in spite of the differences in the two plants, i.e., that STP A has two-step biological removal while STP B has one-step biological removal, in both plants the major removal is determined in the first step. This again may indicate that the relevant step for the mass balance in the STP is a sorption rather than a biodegradation step. Though the concentrations in the respective STPs (A, B, and C) differ considerably, the processes seem to be very similar indeed. As much as 55–75% of the parent compounds are sorbed to sludge. This is obviously a quick process, as it takes place in the first steps in STP A and STP B. About 5–10% of the incoming HHCB is transformed to HHCB-lactone in these STPs. This fact may be of minor relevance for the balancing of these compounds in the STPs, but the concentrations of HHCB-lactone in the discharges of STPs is about 30% of the discharged HHCB. Thus, it will be relevant for the assessment of the fate and effects of HHCB in surface waters (compare Section 2.1.3).

Additionally, it may be interesting to compare the used amounts of fragrances as determined by the concentrations in the inflow of the respective STPs, as the concentrations vary considerably (Table 2.6). It also may be interesting to note that the usage rates from all three cities are comparable, though the socioeconomic data of these cities are very different. Dortmund is an old steel city with high unemployment rates, in contrast to STP A, which has low unemployment and comparatively high incomes. On the other hand, the usages of polycyclic musks in STP B seem to be similar to that of Dortmund, which is not at all reflected by its socioeconomic data. The differences are even larger with respect to the bactericide triclosan (compare Section 2.4).

For comparison, emission data from various sewage treatment plants and surface water samples have been analyzed (see Section 2.1.3 as well as Ref. [47]).

Table 2.6 Comparison of usage of polycyclic musks as determined from wastewater inflow.

		Concentration (ng L^{-1})	Discharge (m^3 d^{-1})	Inhabitants	Annual usage per person (g)
STP A	HHCB	4657	311918	800000	0.66
	AHTN	1086	311918	800000	0.15
STP B	HHCB	2489	130755	320000	0.37
	AHTN	585	130755	320000	0.09
STP C	HHCB	1900	200000	340000	0.42
	AHTN	580	200000	340000	0.13

2.1.2
Polycyclic Musk Fragrances in Diverse Sludge Samples

To gain reliable insight into the emission and processing capabilities of sewage treatment plants in general, the sludge of 20 sewage treatment plants of different sizes was sampled and analyzed for AHTN, HHCB, and HHCB-lactone. The respective samples were extracted and processed as described above. The results exhibited considerable variability (Fig. 2.9). AHTN varied from 500 ng g^{-1} to 7000 ng g^{-1} dry weight, while the concentrations of HHCB ranged from 1000 ng g^{-1} to 15000 ng g^{-1}. The HHCB metabolite, on the other hand, exhibited concentrations of 30 ng g^{-1} to 36000 ng g^{-1}. This high variation of concentrations of the parent compounds is probably partly due to the different rates of mixing between domestic and industrial wastewater in the respective plants. Another important factor might be that mechanisms such as the efficiency of the elimination of total organic carbon (TOC) and humification are different in the diverse STPs. The same effect occurs for other compounds that are used only in domestic applications. In any case, the magnitude of variation is not fully understood in the moment. Possibly, different amounts or differing patterns of fragrances and domestic bactericides are used in the diverse regions. Interestingly, the ratio AHTN : HHCB was more or less constant for all sewage treatment plants (Fig. 2.10). This may indicate that in this part of Germany the consumption pattern of these compounds is rather constant, regardless of whether rural or urban areas are concerned. On the other hand, the relation HHCB : HHCB-lactone varies from 3 to 130. This indicates that degradation processes, especially degradation/transformation efficiency, in the respective STPs differ considerably. Possibly, this is due to the fact that diverse plants with varying hydraulic and sludge retention times were involved in this study. In comparison, in the study on balancing in the STP (Section 2.1.1), relatively low con-

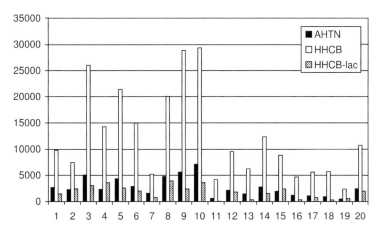

Fig. 2.9 Concentrations (ng g^{-1}) of AHTN, HHCB, and HHCB-lactone in diverse STPs in North Rhine-Westphalia in summer 2002.

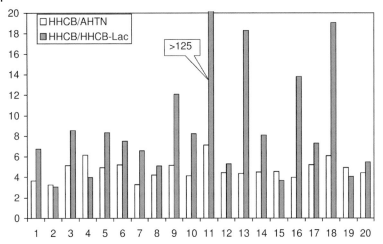

Fig. 2.10 Ratios of HHCB:AHTN and HHCB:HHCB-lactone in sewage sludge of 20 STPs in North Rhine-Westphalia. Value 11 exceeds 125 and thus is not shown fully in this graph.

centrations in sludge (1500 ng g^{-1} AHTN and 3100 ng g^{-1} HHCB) were determined. This is probably due to the fact that this plant receives a considerable amount of its wastewater from industrial sources.

It was not possible to correlate the concentrations of HHCB, AHTN, and HHCB-lactone with the known quotient of industrial/domestic wastewaters, the TOC of the sludges, or any parameters such as working mechanism, e.g., activated sludge or other mechanisms, sludge retention time, phosphate removal, or a combination of these.

At this time it seems that knowledge about the sorption mechanisms of organic micropollutants in STPs is too limited for us to understand quantitatively why or under which circumstances compounds such as HHCB, AHTN, and HHCB-lactone are sorbed to the sludge. On the other hand, this mechanism is important for understanding and predicting the degree of elimination of compounds from the wastewater in an STP. To obtain in-depth insight into this situation, a total balance such as that in STP Dortmund would be needed for each of the respective STPs. This would be prohibitively labor-intensive and costly. Thus, another approach utilizing enantioselective analysis might be used to discriminate between sorption processes and biotransformation processes.

2.1.3
Polycyclic Musk Fragrances in Surface Waters

It has been demonstrated that considerable amounts of polycyclic musk compounds are being used nowadays and that the elimination efficiency of STPs concerning these compounds is limited. The respective plants eliminate some but not all of the respective material. Thus, HHCB, AHTN, and the respective metabolite (HHCB-lactone) (structural formulas in Fig. 2.1) are discharged into rivers as shown by Simonich et al. [25, 26] and Bester [68] (see also Section 2.1.1). This might become an issue, at least in terms of cost, in those areas where the supply of drinking water relies mostly on surface waters, such as the Ruhr area with its approximately 7 million inhabitants. This is important because Seinen et al. [33] found that these compounds exhibit some estrogenic activity. Drinking water in this region is mostly supplied by processing surface water from the Ruhr River. The concentrations of the parent compounds (i.e., HHCB and AHTN but not HHCB-lactone) were monitored with limited spatial resolution some years ago by Eschke et al. [23, 24]. A good overview on HHCB and AHTN in the aquatic environment was given in 1999 by Rimkus [44]. One metabolite of HHCB (i.e., HHCB-lactone) has only recently become available; thus, no quantitative data on HHCB-lactone or its genesis are available from the early studies. This transformation product was identified in 1999 by Franke et al. [35] in surface waters and in 2001 by Kallenborn et al. [34] in fish samples.

It is currently unknown whether or not metabolization/transformation of HHCB and AHTN or other means of elimination exist in riverine ecosystems. Buerge et al. [48] found elimination rate constants of 0.15 d^{-1} and 4.6 d^{-1} for HHCB and AHTN, respectively, in lake water of an oligotrophic Swiss lake; this would lead to half-lives of 4.6 d and 0.15 d for HHCB and AHTN, respectively. However, no transformation products were given in that study. To answer the question of transformation kinetics in riverine and eutrophic lake ecosystems, diverse samples of surface waters from the Ruhr River and several tributaries as well as from STP effluents have been taken in the Ruhr catchment area. The sample extracts were analyzed by normal (non-enantioselective) GC-MS, while some metabolization data were gained by enantioselective gas chromatography, which was used by Franke et al. [35] and Gaterman et al. [31] to determine metabolic processes in fish. The aim of this work was to study input, transportation, and possible elimination by means of metabolization in the river Ruhr.

2.1.3.1 Experimental Methods

The Ruhr River and its main tributaries as well as discharges from significant STPs were sampled within three days in a dry period with no rainfall in September 2002. The river's spring is located in a moderately populated area called Sauerland. After leaving the Sauerland area (station 45, Fig. 2.11), the Ruhr flows through several lakes during its passage through the Ruhr megalopolis. In this area significant amounts of water are extracted for drinking water production. Finally, the remains of the river join the Rhine River in Duisburg. The lakes through which the river passes are shallow, artificial systems (about 1–5 m in depth) and serve to maintain reasonable flow rates for drinking water production plants. They also serve to clarify the river, as particulate matter may settle on the lake's bottom. Additionally, these lakes serve as leisure areas. At the sampled stations (inflow and effluent of the lakes), the water was mostly uniformly mixed, while the lakes themselves most of the time are not. It should be noted that huge efforts are undertaken to keep the Ruhr as clean as possible; thus, major wastewater streams are preferably discharged into the Emscher River, and STPs were put into operation decades ago to protect the Ruhr from the wastewater that is still discharged into it.

The sampling stations of this study are shown in Fig. 2.11, while their characteristics are given in Table 2.7. In most cases the Ruhr was sampled up- and downstream from a possible place of discharge (or tributary) and, wherever possible, the discharge (or tributary) itself was also sampled. Sometimes the plume of the discharge was sampled as well. A plume is the part of the river in which the discharge and the river mix; thus, it does not necessarily represent the river. Plumes sometimes give important information, especially where discharge is not directly accessible.

Water (surface and wastewater) was sampled manually as a grab sample by using 1-L glass bottles with Teflon sealing purchased from Schott (Mainz, Germany). Grab samples are the only appropriate way of sampling if a part of a river of some hundred kilometers in length has to be sampled with a reasonable

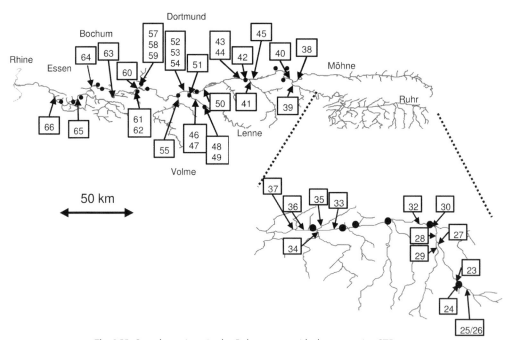

Fig. 2.11 Sample stations in the Ruhr system with the respective STPs (indicated as •). Also indicated are the Rhine River; the Ruhr River tributaries Möhne, Lenne, and Volme; and the cities Essen, Bochum, and Dortmund. Tributary Lenne reaches the Ruhr near sample station 51. (Reprint with permission from [47]).

Table 2.7 Sample station locations, characteristics, and distance from the mouth of the river.

Code	Characteristics	Distance (km)	Date
Blank	Laboratory blank (Millipore water)		17.09.02
23	Effluent of STP Niedersfeld (4000 inhabitants)	211	17.09.02
24	Plume of STP Niedersfeld	211	17.09.02
25/26	Ruhr upstream of STP Niedersfeld	213	17.09.02
27	Ruhr upstream of tributary Neger		17.09.02
28	Ruhr downstream of tributary Neger, concrete plant (no visible effluent)		17.09.02
29	Tributary Neger		17.09.02
30	Ruhr River		17.09.02
31	Field blank		17.09.02
32	Ruhr River		17.09.02
33	Dam/Lock, Ruhr River (Heinrichstal)		17.09.02
34	Tributary Henne		17.09.02
35	Tributary Gebke		17.09.02
36	STP Meschede 1 (downstream of tributaries Henne and Gebke)		17.09.02
37	STP Meschede 2		17.09.02
38	Tributary Möhne	140	17.09.02
39	Downstream of STP Wildshausen-Arnsberg (98 000 inhabitants) upstream of Möhne	145	17.09.02
40	Downstream of tributary Möhne	138	17.09.02
41	Tributary Hönne	117	17.09.02
42	Ruhr downstream of STP Menden-Bösperde (120 000 inhabitants)	116	17.09.02
43	Plume of STP Menden-Bösperde	115	17.09.02
44	Plume of STP Menden-Bösperde	114	17.09.02
45	Ruhr River upstream of tributary Hönne and STP Menden-Bösperde	118	17.09.02
46/47	Tributary Lenne (Motorway) (upstream Hagen STPs)	93	19.09.02
48	Effluent of STP Hagen Fley (17 000 inhabitants)	92	19.09.02
49	Tributary Lenne upstream of STP Hagen Fley (17 000 inhabitants)	94	19.09.02
50	Ruhr River at Schwerte upstream of tributary Lenne and Hagen STPs	95	19.09.02
51	Effluent STP Hagen Boele 17 000 inhabitants (44 000 inhabitants)	92	19.09.02
52/53/54	Ruhr downstream of STP Hagen and tributary Lenne	90	19.09.02
55	Tributary Volme	86	19.09.02
56	Ruhr upstream of STP Ölbachtal, downstream of STP Witten (120 000 inhabitants)	69	19.09.02
57/58/59	Effluent of STP Ölbachtal (160 000 inhabitants)	68	19.09.02
60	Lake Kemnaden, bight into which the effluent of STP Ölbachtal discharges (leisure boat harbor)	67	19.09.02
61	Lake Kemnaden after introduction of the effluent of STP Ölbachtal downstream of station 60	66	19.09.02

Table 2.7 (continued)

Code	Characteristics	Distance (km)	Date
62	Lake Kemnaden after introduction of the effluent of STP Ölbachtal downstream of station 61	65	19.09.02
63	Ruhr downstream of Lake Kemnaden downstream of STP Hattingen (75 000 inhabitants)	60	19.09.02
64	Ruhr downstream of STPs Burgaltendorf, Steele, and Rellinghausen (serving 36 000, 54 000, and 51 000 inhabitants, respectively)	56	19.09.02
65	West end of Lake Baldeney, downstream of STP Kupferdreh (73 000 inhabitants)	37	19.09.02
66	Downstream of STP Kettwig and STP Werden (22 000 and 29 000 inhabitants)	18	19.09.02

Table 2.8 Method quality data for the extraction of musk compounds from aqueous samples.

	Quantifier mass (amu)	Verifier mass (amu)	RT (min)	rr (%)	SD (%)	RSD (%)	LOQ (ng L^{-1})
AHTN	243	258	9.52	78	7	9	1
HHCB	243	258	9.39	75	6	8	3
HHCB-lactone	257	272	14.80	100	23	23	9

RT: retention time; rr: recovery rate; SD: standard deviation; RSD: relative standard deviation; LOQ: limit of quantification determined as 3 times the blank concentration.

spatial resolution. Because surface water is often covered with a biogenic or anthropogenic lipophilic film that may accumulate lipophilic compounds, the sample bottles were passed with a closed stopper through the surface, opened and then closed under the surface of the water at depth of typically 30 cm, and taken out of the water again with a closed stopper. The samples were stored at 4 °C and extracted with 10 mL toluene after adding an aliquot of 100 µL internal standard solution (IS, containing 100 ng D$_{15}$ musk xylene) on the following day. The method validation data are given in Table 2.8. This IS was chosen because it gives an undisturbed signal and musk xylene does not undergo any reaction itself, as this author and Buerge et al. [48] have experienced with deuterated AHTN because it is produced via proton exchange. Thus, deuterated musk xylene is considered a better IS than deuterated AHTN. The same amount of IS was used for all samples. The organic phase was separated from the aqueous one, and the residual water was removed from the organic phase by freezing the samples overnight at −20 °C. The samples were concentrated with a rotary evaporator at 40 °C and 60 mbar to 1 mL. No further cleanup was necessary for quantification with GC-MS. Blank determinations were performed by two different operations. One empty sample bottle was present during the sample trip to control for blanks from the preparation of the bottles and the car (field blank). It was washed with ethyl acetate at the end of the sampling. To for control laboratory and solvent contamination, a 1-L sample of Millipore water was extracted the same way as the surface water samples.

The sample extracts were analyzed by GC-MS (see Section 3.1 and Ref. [49]). However, the photomultiplier of the mass spectrometer was operated at a voltage of 500 V. The mass spectrometer was operated in selected ion monitoring (SIM) mode at 67 ms dwell time.

Enantioselective chromatography was performed utilizing a 25-m column with an inner diameter (i.d.) of 0.25 m, with a film of heptakis-(2,3-di-O-methyl-6-O-*t*-butyldimethyl-silyl)-*β*-cyclodextrin in OV 1701 (obtained as FS-Hydrodex *β*-6TBDM from Macherey-Nagel, Düren, Germany). Film thickness was 0.2–0.3 µm according to the manufacturer. This capillary column was used on a Trace Plus GC-MS system obtained from Thermo-Finnigan (Dreieich, Germany). The transfer line was operated at 210 °C, as the manufacturer suggests a enantioselective column temperature limit of 230 °C. The ion source was kept at 180 °C. The samples were analyzed without further cleanup. The enantioselective separations were performed with the following temperature program: 110 °C (1 min) → 5 °C min^{-1} → 132 °C (140 min) → 1.5 °C min^{-1} → 194 °C (25 min) → 5 °C min^{-1} → 230 °C (10 min). On the first plateau (132 °C) the enantiomers of AHTN as well as the enantiomers and diastereomers of HHCB were separated. On the second plateau (194 °C) separation of the stereoisomers of HHCB-lactone was performed. Helium was used as carrier gas with a flow rate of 0.7 mL min^{-1}.

2.1.3.2 **Results and Discussion**

The methods proved to be sensitive, precise, and robust for these analytes. The recovery rates were 75–100%, the relative standard deviations were 8–23%, and LOQs obtained were three times the field blank (see Table 2.8). Field blank concentrations were 1.1 ng L^{-1} HHCB, 0.37 ng L^{-1} AHTN, and 3.0 ng L^{-1} HHCB-lactone in this experiment, while the laboratory blanks were in the same range but were lower. Thus, the LOQ was lower because of the lower blank values in this experiment than those obtained for the wastewater in Section 2.1.1. Full data are given in Table 2.8.

The concentrations of HHCB, AHTN, and HHCB-lactone found in this study are shown in Fig. 2.12. They ranged from <1.1 ng L^{-1} to 600 ng L^{-1} for HHCB, while they were <0.37 ng L^{-1} to 120 ng L^{-1} for AHTN and <3 ng L^{-1} to 300 ng L^{-1} for HHCB-lactone in the Ruhr and STP effluents. Concentrations exceeding 100 ng L^{-1} were regularly found in the STP's effluents or plumes of effluents entering this river. Concentrations of 100 ng L^{-1} to 600 ng L^{-1} HHCB, 30 ng L^{-1} to 300 ng L^{-1} HHCB-lactone, and 20 ng L^{-1} to 300 ng L^{-1} AHTN were found in the STP effluent samples (stations 23, 24, 36, 43, 57 etc.). Some tribu-

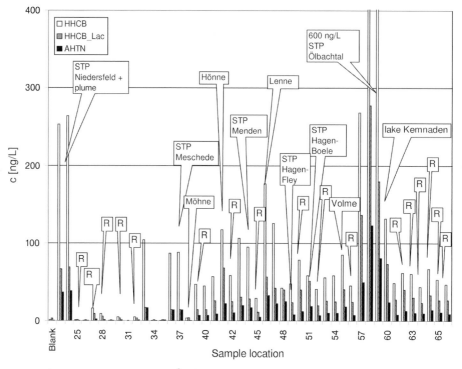

Fig. 2.12 Concentrations (ng L^{-1}) of HHCB, AHTN, and HHCB-lactone in the Ruhr River, some STP effluents, and the tributaries Hönne, Lenne and Volme. Sample 31 is a field blank. R: river sample; STP: STP effluent sample. (Reprint with permission from [47]).

taries such as the Lenne also showed elevated levels. The typical concentrations in the Ruhr in the respective areas from which drinking water is extracted were about 60 ng L^{-1} HHCB, 10 ng L^{-1} AHTN, and 20–30 ng L^{-1} HHCB-lactone. The lowest concentrations (<LOQ) were found near the spring of the Ruhr River. The highest concentrations (several hundred nanograms per liter) were found in the effluent of STPs such as Bochum-Ölbachtal and Niedersfeld (stations 23, 24, and 57/58). However, not in all STPs, e.g., Hagen (station 51) the concentrations of polycyclic musk fragrances in the effluent were higher than in the Ruhr.

Regarding the tributaries Neger (station 29), Henne (station 34), Gebke (station 35), Möhne (station 38), Hönne (station 41), and Volme (station 55), similar or lower concentrations than in the Ruhr were determined. On the other hand, samples from the tributary Lenne (stations 46–49), which may be connected with a small STP, exhibited elevated levels of these analytes.

In the main part of the river, i.e., the most populated parts (stations 50–66), the concentrations were about 60 ng L^{-1} HHCB, about 30 ng L^{-1} HHCB-lactone, and about 10 ng L^{-1} AHTN without significant changes. No significant input can be detected, e.g., via increasing concentrations, from these data downstream of station 61, although several small STPs discharge into this part of the river (Table 2.7). Upstream of station 61 the wastewater of more than 1.4 million inhabitants is discharged into the Ruhr, while between stations 61 and 66 the treated wastewater of 340 000 inhabitants is discharged into the river. Thus, in the area near the mouth of the river, the discharge is less than 25% of the amounts discharged in the upper parts of the river. This would be near the RSD of the analytical method. The fact that these STPs are spread over some distance may be the reason that these STPs were not identified as significant sources of discharge. Additionally, it should be noted that the main flow of wastewater of the northwestern part of the Ruhr megalopolis (including most parts of the cities of Essen and Bochum, etc.) is directed to the Emscher River.

Samples for comparison were taken from the Rhine River near Düsseldorf and from the Lippe River in the vicinity of the Ruhr megalopolis. In the Rhine, concentrations of around 20 ng L^{-1} HHCB, 4 ng L^{-1} AHTN, and 20 ng L^{-1} HHCB-lactone were determined. The samples from the Lippe gave concentrations of 75 ng L^{-1} HHCB, 12 ng L^{-1} AHTN, and 30 ng L^{-1} HHCB-lactone. These concentrations agree well with those of Dsikowitzky et al. [22], who found concentrations of about 100 ng L^{-1} for HHCB and 50 ng L^{-1} for AHTN in a similar part of the river in samples taken in the winter of 2000. In 1995 the concentrations in the Elbe River were about 100 ng L^{-1}, with concentrations of HHCB and AHTN similar to those shown in 1998 by Bester et al. [39] and in 1999 by Franke et al. [35]. Nowadays, HHCB dominates significantly over AHTN and HHCB-lactone in all samples taken; thus, a change in application patterns is obvious.

With regard to temporal changes, the Ruhr and some tributaries were sampled for HHCB and AHTN in 1993 by Eschke et al. [23, 24]. At that time, neither analysis of the metabolite HHCB-lactone nor enantioselective analysis was possible. While high concentrations, e.g., in the effluent of STP Ölbachtal, have not changed significantly during the past decade, the typical river concentrations seem to have changed: they were reported to be 500 ng L^{-1} for HHCB and 300 ng L^{-1} for AHTN in 1993, while now they are 60 ng L^{-1} and 10 ng L^{-1}, respectively. Thus, it seems that river background values have decreased considerably. The STP situation in this region has changed, as a multitude of older and smaller STPs have been decommissioned, and wastewater is more often treated in larger, more centralized facilities that use current technologies.

Additionally, it should be noted that the water mass flow is not constant in rivers such as the Ruhr. During this new study, the water flow in the Ruhr was less than average. The data of the older studies are not available. Thus, it seems that the riverine values have decreased considerably. This finding may be attributed to the efforts of manufacturers of laundry detergents to replace polycyclic musk compounds and nitroaromatics.

In 1999 Heberer et al. [50] published data on HHCB and AHTN concentrations in water samples from surface and discharge waters in Berlin, Germany. In that study the concentrations ranged from 20 ng L^{-1} to 12 500 ng L^{-1} for HHCB and from 30 ng L^{-1} to 5800 ng L^{-1} for AHTN. Although Berlin's surface waters are continuously heavily polluted by STP discharges, the highest concentrations were found in STP effluent samples. The typical river concentrations were similar to those obtained in this study. For HHCB-lactone, there were no quantitative data published.

Transformation Data

To gain some insight into the persistence and possible transformation of HHCB, the ratio of HHCB-lactone versus HHCB was calculated from all current Ruhr samples with significantly elevated concentrations. The respective graph is shown in Fig. 2.13. It should be noted that the technical product of HHCB contains about 10% HHCB, which is also found in the STP influent as shown in Fig. 2.14. The enantiomeric ratio of lactone (4S,7S vs. 4R,7R as well as 4S7R vs. 4R,7S) HHCB-lactone in the technical product is 1.0 as well as in raw wastewater (inflow of the STPs).

It should be noted that the diverse STP discharges that were sampled gave differing ratios of HHCB-lactone to HHCB. It seems that each STP exhibits its own metabolic profile. In STP effluent samples the ratio ranged from 0.17 (no. 36) to 0.95 (no. 48, 51). A comparison of the STP technologies which are used in the STPs in the respective sampling series is given in Table 2.9. It seems that each STP has its individual emission pattern that can hardly be obtained from theoretical data.

In samples from the Ruhr with high concentrations (downstream from Schwerte [station 50]), the ratio HHCB-lactone:HHCB was constantly 0.5 in all riverine samples during the passage of the lakes until station 66 was reached. However, some tributaries, e.g., Lenne, exhibited slightly lower ratios. The distance from stations 57 to 66 is about 50 km, and the river passes several lakes,

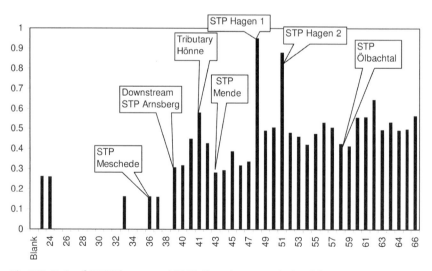

Fig. 2.13 Ratio of HHCB-lactone vs. HHCB. No ratios were calculated from values near the blank concentration. (Reprint with permission from [47]).

Table 2.9 Main STPs in the sampling area.

STP	No.	Effluent (m³ d⁻¹)	Type	hRT (h)	BOD re-moval [a] (%)	COD removal [a] (%)	Nitri-fication [a]	Denitri-fication	Ratio mun (%)	C HHCB (ng L⁻¹)	C HHCB-lactone (ng L⁻¹)	ER Lac (SS/RR)	C AHTN (ng L⁻¹)
Niedersfeld	24	3800	AS, clari	14	92	86	no [b]	no	100	264	69	0.883	38
Arnsberg-Wildhausen	n.a.	20800	AS, clari	15	97	89	yes	yes	74	n.a.	n.a.	n.a.	n.a.
Menden Bösperde	43	38900	AS, trick-ling filter	3.5	92	83	no	no	75	106	30	0.930	19
Hagen–Fley	48	12900	AS, clari	3.3	98	78	no	no	100	42	40	n.a.	25
Hagen–Boele	51	Not yet fully established							55	58	51	n.a.	18
Bochum-Ölbachtal (mean)	57	70300	AS, clari	22	98	93	yes	yes	63	451	198	0.783	84

a) Data obtained from the waterworks company.
b) Nitrification does not exist in the respective plant.
No.: sample number; BOD: biological oxygen demand; COD: chemical oxygen demand; C: concentration; AS: activated sludge; hRT: hydraulic residence time; n.a.: data not available; Ratio mun: ratio municipal vs. industrial contribution; ER lac: enantiomeric ratio for HHCB-lactone; SS/RR: enantiomeric description; clari: clarifier.

including Lake Kemnaden (3×10^6 m³), Lake Baldeney (8.3×10^6 m³), and Lake Kettwig (1.5×10^6 m³). Therefore, it was suggested that if degradation of HHCB in the river took place, this would be the most probable place it could be determined. However, in this region the ratio of HHCB versus HHCB-lactone did not change significantly, thus indicating a rather conservative transport of these compounds to the mouth of the river without significant transformation from HHCB to HHCB-lactone. This is supported by the fact that the concentrations also remain constant. These ratios have been analyzed with about 15% uncertainty. This means that the transformation of HHCB to HHCB-lactone in the river was less than 15% of the riverine inventory of HHCB. On the other hand, in STPs HHCB can be transformed to some degree into the lactone (Fig. 2.1) [68]. In the upstream part of the Ruhr River, the concentrations were too low to lead to clear conclusions. Additionally, a variety of discharges from small STPs made it difficult to interpret these results.

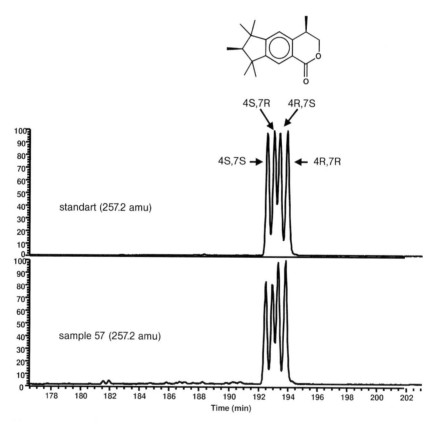

Fig. 2.14 Enantioselective separations of the enantiomers of HHCB-lactone in an STP effluent sample. Also indicated is 4*R*,7*S* HHCB-lactone (Elution order following Meyer [51]). (Reprint with permission from [47]).

A comparison of the Ruhr data with those obtained from similar experiments on the Rhine gave a HHCB-lactone vs. HHCB ratio of 1.0, while for the Lippe River a ratio of 0.4 was determined. In the Elbe, HHCB and HHCB-lactone ratios were tentatively determined by Franke et al. [35] and Meyer [51]. In this river the concentrations of HHCB-lactone were 30% of those found for HHCB. It may be that the different ratios of HHCB vs. HHCB-lactone in these rivers (Ruhr, Elbe, Rhine, Lippe) are due to the different STP technologies used. This is also indicated in this set of data, as diverse STPs do not emit the same ratios.

Currently it is not possible to correlate STP emission concentrations or the respective ratios with the technology applied in the respective plant. This has also been a difficulty for other authors [26] as well as in the understanding of another dataset on polycyclic musks in sludge. With regard to AHTN, there is no significant change in concentrations in the river from stations 50 to 66. In this part of the river, AHTN concentrations are 19% of HHCB concentrations, with 2% standard deviation. A slight increase would be expected, as the STPs in that part of the river could contribute up to 25% of the total load. This still could be due to the uncertainty of measurement. In any case, if degradation of AHTN does occur, the estimation based on these data is less than 25% during the passage of the lower part of the river, even though the sampling took place in bright, warm, summer weather in a dry period without any rainfall, with conditions favorable to photo- and biodegradation.

Enantioselective Determination for the Verification of Transformation Data
The enantioselective measurements, on the other hand, are more precise than non-enantioselective determination, as they are relative determinations to the respective enantiomer and thus operate on an excellent intrinsic internal standard. In Fig. 2.14 the separation of the stereoisomers of HHCB-lactone is

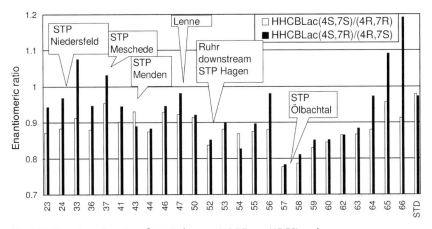

Fig. 2.15 Enantiomeric ratios of HHCB-lactone (4S,7S) vs. (4R,7R) and (4S,7R) vs. (4R,7S) in selected Ruhr water as well as STP effluent (STD: standard solution). (Reprint with permission from [47]).

shown, while the elution order determined by Meyer [51] is used. Because HHCB-lactone has two chiral centers, four peaks are resolved. The distribution of enantiomeric ratios of HHCB-lactone in the samples is shown in Fig. 2.15. These analyses were performed to study whether in all STPs the same microorganisms (enzymes) performed the transformation, or whether diverse biological processes led to different enantiomeric ratios.

The enantiomeric ratios (ERs) measured for HHCB-lactone enantiomers (4S,7S) vs. (4R,7R) in the standard solutions were 0.98, with 0.015 (1.5%) RSD, similar to those determined for wastewater inflow of STPs. The samples obtained from the Ruhr upstream of Schwerte (station 50) exhibited ERs of about 0.9, while the samples from the effluent of STP Ölbachtal exhibited an ER of about 0.8 (stations 57–59). In some, though not essentially all, STPs HHCB may be transformed to some extent to HHCB-lactone by biological processes. Thus, verification on the hypothesized biotransformation is achieved.

Considering the ratio of HHCB-lactone enantiomers (4S,7R) vs. (4R,7S), the standard exhibited an ER of 0.97 (1.8% RSD) as well as raw wastewater (inflow in the STPs), while the Ruhr River samples had ERs of 0.9 to 1.0 upstream of station 57. The samples obtained from the Ruhr upstream of Schwerte (station 50) exhibited ERs of 0.9. STP Ölbachtal emitted HHCB-lactone with an ER of 0.8; thus, some biotransformation process of HHCB was ongoing during the sewage treatment process. Downstream of this situation, another process seems to become important, as ratios changed to even 1.2.

The finding that in most cases ERs are smaller than 1 (only in the case of (4S,7R) vs. (4R,7S) is the ER higher than 1) may indicate that there are different biological processes, e.g., diverse enzymes, that may have some relevance in samples 33, 65, and 66. It should be noted, though, that overall only a small percentage (~7%) of the HHCB in the river may be transformed to HHCB-lactone.

The enantiomeric ratios of AHTN and the respective stereoisomers of HHCB have also been determined. No significant changes in those have been detected. For AHTN this is probably due simply to low microbial degradation, while for HHCB two factors may be involved:

1. Because a major fraction of HHCB-lactone is formed in the STP, changes are more easily detected than for HHCB. If for instance 5% of the HHCB were transformed to HHCB-lactone, a maximum change in ERs for HHCB would be 5%, which would be very hard to detect. This could typically represent, e.g., 50% of the HHCB-lactone (the residue could typically originate from product impurities). Therefore, changes in ERs of transformation products are much easier to detect.

2. The enantioselective separation of HHCB may sometimes face co-elution problems with the other impurities of the product. Again, the precision here is not as good as for the determination of enantiomeric ratios of HHCB-lactone.

The analysis of enantiomer composition indicates the biotransformation of HHCB to HHCB-lactone in STPs such as Bochum-Ölbachtal. However, there are only small indications for biotransformation in the river and the lakes. The situation in other rivers, e.g., in the Rhine, that exhibited HHCB to HHCB-lactone ratios of 1:1 may be different. The concentrations in the Rhine measured in this study were too low (< 30 ng L^{-1}) for enantioselective analysis.

The typical river levels for HHCB, HHCB-lactone, and AHTN in the Ruhr are about 60, 25, and 10 ng L^{-1}, respectively. This means that the Ruhr, with a total flow of about 2200 million m^3 water annually, transports about 130–350 kg HHCB, 55–150 kg HHCB-lactone, and 22–60 kg AHTN (depending on the annual transport patterns) into the Rhine each year.

On the other hand, immission scenarios for the river Ruhr of 0.1–0.13 g HHCB per person annually (pp/a), 0.014–0.04 g AHTN pp/a, and 0.04–0.07 g HHCB-lactone pp/a have been established from outflow data of major STPs in this region (Fig. 2.12). These are emission rates to the river after the STP process, including the elimination and transformation rates of the STPs. This leads to the observation that 230–310 kg HHCB, 90–270 kg HHCB-lactone, and 36–70 kg AHTN are discharged to the river via STP effluents. This mass balance approach gives as result that all the HHCB, HHCB- lactone, and AHTN that is discharged into the Ruhr is transported to its mouth, and no significant degradation can be observed during the passage of the river in summertime. It seems that the Ruhr is in a kind of steady state, i.e., what is imitted to the Ruhr is discharged to the Rhine without significant sedimentation or biodegradation processes. The Rhine holds concentrations of about 30 ng L^{-1} HHCB, 5 ng L^{-1} AHTN, and 20 ng L^{-1} HHCB-lactone in the Rhine/Ruhr area. Because the water flow of this river is about 7.4×10^{10} m^3 year^{-1} in this river [52], about 2.2 t HHCB, 0.4 t AHTN, and 1.5 t HHCB-lactone are transported annually with the Rhine via the Netherlands to the North Sea.

The concentrations of HHCB and even more so of AHTN in the Ruhr decreased in the last decade. However, the effluents of larger STPs still exhibit high concentrations (100–600 ng L^{-1} HHCB) similar to those in 1993. The efficiency of the smaller STPs may have increased, as there has been considerable reconstruction of STPs in this area.

A considerable amount of HHCB is transformed into HHCB-lactone in the STPs, which should also be taken into account when performing assessments of polycyclic musk fragrances. Because the transformation process is enantioselective in some STPs, the degradation in these plants is probably due to enzymatic transformation, not to abiotic oxidation. No indication for removal or biotransformation of these polycyclic musks during the passage of the Ruhr was detected in this study, though several lakes could serve as a sedimentation sink. All possible changes such as sorption to sediment, volatilization, biodegradation, and photodegradation should include changes of ratios of HHCB versus HHCB-lactone, because all physicochemical parameters, such as log K_{ow}, differ significantly. The log K_{ow} of HHCB-lactone is 4.71, while it is above 6 for HHCB [53]. It cannot be ruled out, however, that several processes add up so that no changes in ratios appear. This seems to be very improbable at the current state of results, though. It thus seems reasonable to assume that degradation takes place neither in the shallow lakes nor in rivers such as the Ruhr. Because of high standards in water purification, the adverse effects of these compounds on human health will probably be low. On the other hand, the largest fraction of these compounds reaches human bodies via the skin from perfumes or treated clothing [32]. Additionally, it should be noted that the consumer has to pay considerably for the water purification. All three compounds add to the load of xenobiotics that wildlife has to cope with when living in industrialized riverine systems. An additional issue arises when fish are caught in these rivers and consumed afterwards. The presence and fate of HHCB-lactone should be included in all assessments concerning HHCB. In particular, toxicity data on HHCB-lactone are needed for further assessment.

2.1.4
Polycyclic Musk Fragrances in the North Sea

Within a non-target screening approach of water samples of the Elbe River estuary, the two polycyclic musk fragrances HHCB and AHTN were identified. These compounds were found for the first time in the aquatic environment by Eschke et al. [24]. Lagois [54] found concentrations from < 20 ng L^{-1} to 100 ng L^{-1} in water samples taken from the upper part of the Elbe in Saxony, which may indicate that these compounds enter the rivers via the STPs. Franke et al. [55] also identified HHCB and one isomer, probably AHTN, in all samples taken from an upstream region of the Elbe. Eschke et al. [23, 24] analyzed water samples from the Ruhr (a tributary of the Rhine) and found even higher concentrations than Lagois [54] (about 200–370 ng L^{-1}). In these earlier studies it was assumed that these compounds enter the aquatic environment via the STPs as laundry detergent and cosmetic fragrances and are then diluted in the rivers [24, 56] (compare Sections 2.1.1–2.1.3). About 4300 t per annum of these fragrances were produced worldwide in 1987 [57]. These compounds are assumed to replace nitroaromatic musks such as musk xylene, which was partially phased out because of its toxicity and persistence [56, 58]. Because data on the presence of polycyclic musk fragrances in the environment are rare at the moment, it was the aim of this study to identify and quantify these highly lipophilic compounds within water samples from the German Bight of the North Sea.

Because the concentrations in the North Sea are essentially much lower than in riverine systems, different analytical methods are needed. The procedures for seawater are discussed in Section 3.2. In this case a sectorfield mass spectrometer was applied, with which elemental composition also could be assessed. Both HHCB and AHTN, which possess the same theoretical mass, were clearly identified within this study using the EI mass spectrum for interpretation. Mass spectral data were obtained from the base peak (bp) for HHCB bp = M–Me ($C_{17}H_{23}O$), with the theoretical mass of 243.175 compared to the measured mass 243.187, while the mass determined for AHTN was 243.193. The respective mass spectra and retention times were compared to those obtained from injecting a solution of the pure standard. Because both mass spectra and retention times of HHCB and AHTN in the sample extract and in the standard solution were identical, the presence of these compounds in the water samples of the river Elbe estuary and the German Bight was confirmed (see Figs. 2.16 and 2.17).

Recovery rates (rr) were analyzed by extracting 1 L spiked water (2.5–250 ng L^{-1} HHCB and 2–200 ng L^{-1} AHTN, respectively), and they represent the extraction only. The concentrations discussed below are not corrected for recovery rates, as it does not represent the whole procedure. The values of the rr as well as the limit of quantification (LOQ) (obtained from the calibration function) and blanks are given in Table 2.10. LOQs were determined by the following routine. During the quantification experiment, a quantitative calibration of the mass spectrometer was performed. Within this experiment 20 pg HHCB and 15 pg AHTN injected into the GC-MS were reproducibly quantified, while lower concentrations varied because of small signal-to-noise ratios. Considering the method, LOQs were calculated. Further details are given in Section 3.2.1.

In June 1990 and from June to July 1995, several 100-L water samples were taken in the German Bight within the German monitoring program (BLMP) of the North Sea in order to quantify the concentrations of diverse "classical" pollutants. These samples were also used for analyzing the concentrations of HHCB and AHTN. At all seven stations investigated herein, both compounds were

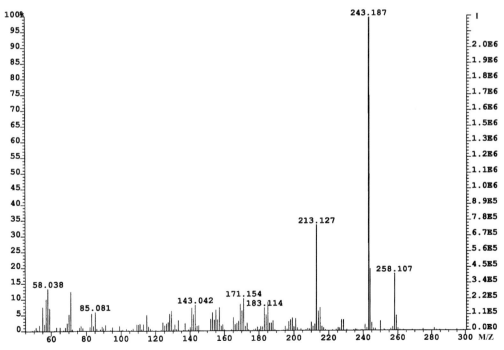

Fig. 2.16 Mass spectrum of HHCB obtained from a water sample from the river Elbe estuary. I = relative intensity. (Reprint with permission from [273]).

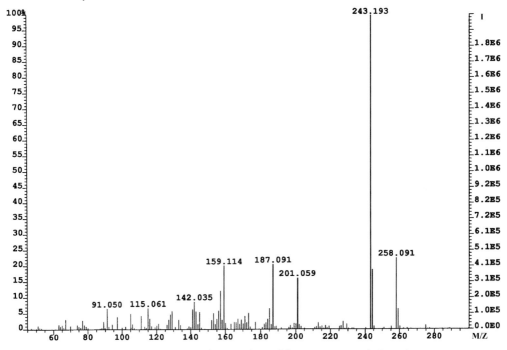

Fig. 2.17 Mass spectrum of AHTN obtained from a water sample from the river Elbe estuary. I = relative intensity. (Reprint with permission from [273]).

Table 2.10 Limits of quantification (LOQ) and blanks as obtained by 100-L samples, while recovery rates (rr) were measured by extraction of 1-L samples. All data are given in virtual water concentrations for polycyclic musk compounds from 100 L seawater.

	LOQ (ng L^{-1})	rr (%)	Blank (ng L^{-1})
HHCB	0.04	87	< LOQ
AHTN	0.03	66	< LOQ

present. Ion-fragment chromatograms (SIM) obtained from an extract of a water sample and a standard solution are shown in Fig. 2.18. The concentrations were calculated using D$_8$ naphthalene as an internal standard. No quantification is available for the estuarine sample from 1990, as this was used for identification.

The spatial distribution of the respective compounds is shown in Fig. 2.19, while the concentrations are summarized in Table 2.11. The highest concentrations were observed in the Elbe River estuary, where values of 95 ng L^{-1} HHCB and 67 ng L^{-1} AHTN were encountered. Decreasing values towards the northern and northwestern stations in the German Bight suggest increasing dilution of these compounds in marine waters. The respective concentrations at the north-

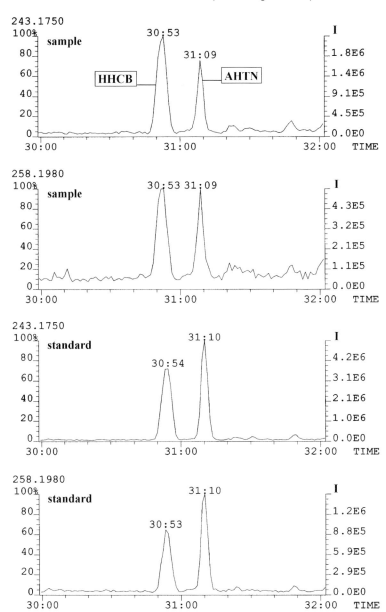

Fig. 2.18 Ion-fragment chromatograms (analyzed in SIM mode) of masses 258.198 and 243.175, significant for both HHCB and AHTN (upper two chromatograms: sample from station 4 taken in the north of the sampling area; lower two chromatograms: standard; I=intensity).
(Reprint with permission from [273]).

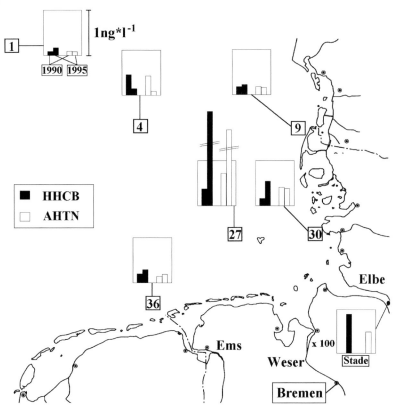

Fig. 2.19 Distribution of HHCB and AHTN in the water of the German Bight of the North Sea. Data are given for 1990 and 1995. Maximum values are 1 ng L^{-1} within the fields. The rivers Elbe, Weser, and Ems are indicated as well as the city of Bremen and the sample station Stade. (Reprint with permission from [273]).

western edge of the sampling area are 0.09 ng L^{-1} and 0.08 ng L^{-1} (compare Table 2.11). It is interesting to note that there is a sharp decline from the estuary to the marine sampling stations that was not observed for other compounds such as triazines, benzothiazoles, anilines, etc. (see Sections 2.4.6, 2.5.3 and 2.6.1). However, this sharp drop could not be reproduced in recent studies (see Section 2.4.6). No further clues concerning inputs and sources of these compounds can be drawn from their geographical distribution.

The concentrations of HHCB and AHTN found at the seawater border of the estuary of the Elbe (sample station Stade) are of the same order of magnitude as the most polluted sites at the upstream part of the river (100 ng L^{-1}) [54] and in the Swiss river Glatt 3 km from an STP (136 ng L^{-1} HHCB and 75 ng L^{-1} AHTN) [59]. The values found in the present study are lower than those obtained by Eschke et al. [23] or even those determined in Section 2.1.3 for the river Ruhr (about 370 ng L^{-1} HHCB and 200 ng L^{-1} AHTN), which is supposed to be highly

Table 2.11 Concentrations (ng L^{-1}) of polycyclic musk compounds in the water of the German Bight of the North Sea in 1990 and 1995 and the river Elbe in 1995.

	HHCB 1990	HHCB 1995	AHTN 1990	AHTN 1995
Sample station				
1	0.09	0.17	0.09	0.08
4	0.45	0.15	0.43	0.08
9	0.18	0.22	0.18	0.16
27	0.88	4.8	0.94	2.6
30	0.16	0.55	0.41	0.38
36	0.20	0.30	0.14	0.19
Stade/Elbe	Not quantified	95	Not quantified	67

polluted by STPs. The concentrations in the estuary of the Elbe River are comparatively high when it is taken into account that the sampling station Stade (between the city of Hamburg and the North Sea) is normally considered to represent water of the Elbe that is already diluted by seawater to some extent. The STP in the city of Hamburg, located about 30 km upstream of the sampling station, discharges the treated sewage water of about 1.5 million inhabitants and may be causing a notable input of these polycyclic musk fragrances, though no data are available on the outflow of these compounds from this specific plant.

The sharp drop in concentrations in the mouth of the estuary (Stade in comparison to station 30) may be attributed to the sedimentation zone of this estuary, as the polycyclic musk fragrances have high *n*-octanol–water partition coefficients (log K_{ow}) of 5 to 6 [24] and compounds with those characteristics are known to sorb onto particles with a high content of organic carbon, like those in the sedimentation zone. This may lead to high concentrations in the sediment in this area. However, this hypothesis could not be verified yet. It is also possible that the high concentration in "Stade" in this sampling series is a mere sampling artifact, possibly due to sampling in a wastewater plume.

Interestingly, there seems to have been an increase in concentrations of HHCB from 1990 to 1995 at most stations. This result contradicts the general assumption that because of political and economic changes in the Elbe catchment's basin, pollution in the Elbe and inputs of pollutants to the North Sea would decrease, as actually was observed with regard to some other pollutants [60]. It may be hypothesized that the input of polycyclic musk compounds is not decreasing but increasing because of their growing production numbers from 1990 to 1995. This observation has to be verified in further, more-detailed studies, as other effects, e.g., hydrodynamic and meteorological changes or seasonal effects, may have influenced the respective concentrations.

The concentrations found for both HHCB and AHTN in the water of the German Bight are of the same order of magnitude (see Table 4.3) as those measured for α- and γ-HCH [61], nitrobenzene [62], and thiophosphates such as *O,O,O*-trimethylthiophosphate and *O,O,S*-trimethyldithiophosphate [63]. They are considerably lower than those found for triazine herbicides in coastal waters [64]. On the other hand, the input into the North Sea via the Elbe seems to be of the same order of magnitude as methylthiobenzothiazole [60], being one order of magnitude less than the triazine herbicides but still exceeding the other pollutants mentioned. Obviously, these compounds do not behave conservatively in the estuary, i.e., they are not only diluted in marine water but also are, e.g., sedimenting adsorbed to particles or being transformed to some extent. It may be interesting to note that the concentrations of polycyclic musk fragrances described in this study exceed those of the nitroaromatic musk perfumes musk ketone and musk xylene, which were described by Gatermann et al. [62], by a factor of 3 to 4 (see Table 4.3), reflecting the increasing production and usage of HHCB and AHTN [57].

The polycyclic musk fragrances HHCB and AHTN are highly lipophilic and thus bioaccumulate to a high extent and by diverse paths [23, 24, 56, 65]. The high log K_{ow} may be the reason for high concentrations of these perfumes in the lipids of eels and pikeperches from, e.g., the Elbe River [44]. The respective log K_{ow} values are of the same order of magnitude as those of the lipophilic PCBs. The concentrations of polycyclic musk fragrances investigated in seawater in this study exceed those of PCBs by a factor of about 3. This may be attributed to the fact that the usage of PCBs has been phased out in Europe (these compounds may be used only in closed systems), while the open application of polycyclic musk fragrances, most of which are probably discharged into the environment, has been growing continuously [57]. For an assessment of these compounds, it must be noted that there are no ecotoxicological and environmental data on these compounds apart from their high log K_{ow}s and high bioaccumulation. Few other toxicological data on this group of substances are available at the present. One polycyclic fragrance compound that has already been phased out, i.e., Versalide® (7-acetyl-1,1,4,4-tetra-methyl-6-ethyl-tetrahydronaphthalin), was found to be neurotoxic [66]. At the present state of the art, it is completely unknown whether or not the concentrations of HHCB and AHTN found in water of the North Sea have any effect on marine wildlife. However, the observed concentrations of highly bioaccumulating compounds should be considered a cause for concern.

2.1.5
OTNE and Other Fragrances in the Environment

OTNE ([1,2,3,4,5,6,7,8-octahydro-2,3,8,8-tetramethylnaphthalen-2yl] ethan-1-one) has been among the most popular compounds in fragrancy in the last few years (structural formula given in Fig. 2.20). It is marketed as Iso E Super®, with 2500–3000 tonnes annually being sold, while, e.g., HHCB (Galaxolide®) is sold at a volume of 7000–8000 t/a [67]. Though the environmental behavior of fragrance compounds has been debated in the cases of musky compounds such as HHCB, AHTN, musk xylene, musk ketone, etc. [23–25], little is known about the fruitier synthetic fragrance compounds. Of the musk compounds it is now well known that they are not mineralized in sewage treatment processes and that sorption

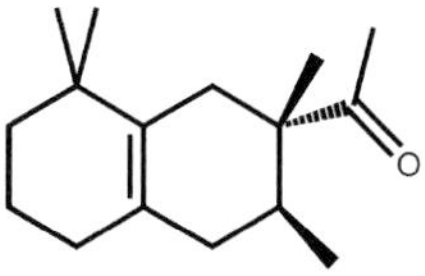

Fig. 2.20 Structural formula of OTNE [71].

is their main elimination path in sewage treatment, although transformation to other compounds may occur [47, 68]. On the other hand, in the cases of musk xylene and HHCB, metabolites that are formed in the STP process are relevant emission parameters of current STPs. OTNE has escaped such attention until to now. Some data on OTNE in STPs and soil and its volatility have been made available to the public by Simonich [26], Difrancesco [70], and Aschman [69]. Basically, OTNE was reported to be not readily biodegradable in OECD tests, and concentrations of 3.6 μg L^{-1} were detected in wastewater from the U.S., while in Europe the concentrations were 9.0 μg L^{-1}, i.e., in the U.S. the concentrations were lower than those of AHTN and HHCB, while in Europe these concentrations were equal [26]. The concentrations of OTNE in the effluents ranged from 0.03 μg L^{-1} to $\sim$ 4 μg L^{-1}, with the European samples being the highest. In that study the technologies currently applied most often, i.e., activated sludge and settling, resulted in removal rates of 92% and 29%, respectively [26]. In another study the fate of OTNE in sludge-amended soils was described. Concentrations of 7–30 μg g^{-1} OTNE in dry sludge were determined in sludges from the U.S. In the soil, half-lives of 30–100 d were determined for OTNE [70].

Additionally, OTNE has peculiar properties. It was previously on the market as the technical mixture Iso E Super®, which was well received by the consumer. It was only recently discovered that the compound that causes the (wanted) sensoric effect in that mixture is a byproduct that became available as a pure compound just a few years ago and is currently sold as Iso E Super Plus® [67, 71]. It is thus of not only environmental but also economic relevance to have published data on the fate of the main constituents of Iso E Super. Therefore, OTNE was analyzed in this study in the influent and the effluent of some STPs as well as the surface water of the Ruhr from which about 5 million inhabitants of the Ruhr megalopolis are served with drinking water.

2.1.5.1 Methods

OTNE was studied during a dry period in sewage treatment plant C (near the city of Dortmund), which was described in Section 2.1.1. The Ruhr River and its main tributaries, as well as discharges from significant STPs, were sampled within three days during a period with no rainfall in September 2002 (see Section 2.1.3). The unfiltered wastewater and surface water samples were stored at 4 °C and extracted within 24 h for 20 min with 10 mL toluene after adding an aliquot of 100 μL internal standard solution (with 100 ng D$_{15}$ musk xylene). The method validation data are given in Table 2.12. Details of this method are discussed in Section 3.1.

Table 2.12 Method validation data. The limit of determination (LOD) was determined from the recovery experiments, while the LOQ considers the field blank of 70 ng L^{-1}.

	Recovery rate (%)	SD (%)	LOD (ng L^{-1})	LOQ (ng L^{-1})
OTNE	122	13	10	100

Blanks were obtained by two different operations. One empty sample bottle was present during the sample trip to control blanks from the preparation of the bottles and the car (field blank). It was washed with ethyl acetate at the end of the sampling. To control for laboratory and solvent contamination, a 1-L sample of Millipore water was extracted the same way as the surface water samples.

Final analysis was performed by GC-MS (Thermo-Finnigan DSQ) equipped with a PTV injector and an AS 2000 autosampler. The PTV (BEST PTV, Thermo-Finnigan, with 4-μL injection volume) was operated in PTV splitless mode applying the following temperature program: $90\,°C$ (0.1 s) $\rightarrow 12\,°C\ s^{-1} \rightarrow 280\,°C$ (1 min) (cleaning phase; open valve). The GC separation was performed utilizing a DB-5MS column (J&W Scientific, Folsom, CA, USA) (length: 15 m; i.d.: 0.25 mm; film: 0.25 μm) and a temperature program of $100\,°C$ (1 min) $\rightarrow 30\,°C\ min^{-1} \rightarrow 130\,°C \rightarrow 8\,°C\ min^{-1} \rightarrow 220\,°C \rightarrow 30\,°C\ min^{-1} \rightarrow 280\,°C$ with $1\ mL\ min^{-1}$ He as carrier gas. The transfer line was held at $250\,°C$, which is sufficient to transfer all compounds from the gas chromatograph into the MS as the vacuum builds up in the transfer line. The ion source was operated at $250\,°C$. For quantification, the mass fragments 191.0 and 219.0 were analyzed in selected ion monitoring mode (SIM) with 40 ms dwell time and the multiplier set to 1089 V.

This procedure was validated with pure water and gave recoveries of about 120 (± 13)% when the internal standard was used. Full quality data of the method obtained from three replica extractions at five different concentrations are given in Table 2.13. The LOD was established by the recovery experiments to be 10 ng L^{-1}. The field blank during the sampling of surface water was observed to be 70 ng L^{-1}. Application of the 3 σ criterion (three times standard deviation of the field blank added to this value) results in a field-based LOQ of 100 ng L^{-1}.

The chromatogram of mass fragment 191.0 atomic mass units (amu) for the base peak (M -C$_2$H$_3$O) of OTNE was used for quantification, while mass 219.0 amu (M -Me) was used for verification. The internal standard D$_{15}$ musk xylene was analyzed at 294 amu (M -CD$_3$) and 312 amu (M), respectively. The calibration was performed as a seven-step internal standard calibration with linear regression of 1–1000 ng mL^{-1} (extract concentrations) and 1/x weighting.

2.1.5.2 Results and Discussion

Technical OTNE (Iso E Super) exhibits a pattern of several peaks with similar mass spectra. The mass spectrum of the main constituent, as obtained from a sample, is shown in Fig. 2.21. This spectrum is identical to those obtained from a technical standard of Iso E Super. The suggested fragmentation pattern was verified by exact mass determination for mass fragment 191, which matched with 1.5 ppm aberration. A comparison of retention times and peak pattern is shown in Fig. 2.22. Iso E Super was thus identified in STP effluents as well as in surface waters (Fig. 2.23). All concentrations in STP influents and effluents were significantly above the limit of quantification.

To study elimination efficiencies in STPs, the concentrations of OTNE were determined in a four-day experiment in influent and effluent water from 24-h composite samples (Table 2.13). In these samples the concentrations were so

Table 2.13 Concentrations of OTNE in STPs in and effluent obtained by 24-h composite samples.

	Influent (ng L^{-1})	Effluent (ng L^{-1})	Breakthrough (%)	Elimination (%)	Sludge (ng g^{-1})
1	4739	1809	38	62	3600
2	3576	1560	44	56	2900
3	4528	840	19	81	4500
4	3494	1378	39	61	3800
Average	4084	1397	35	65	3700
SD	641	411	11	11	690

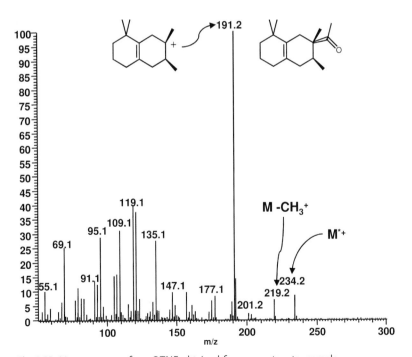

Fig. 2.21 Mass spectrum from OTNE obtained from a wastewater sample.

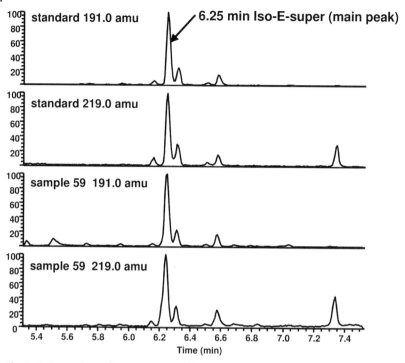

Fig. 2.22 Comparison of retention times of Iso E super from a standard and from station 59 (STP effluent).

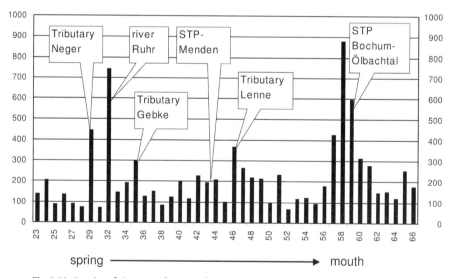

Fig. 2.23 Results of the quantification of Iso E Super in surface waters and STP effluents. All concentrations given in ng L^{-1}.

high that both mass fragment were used for quantification. The concentrations in influents exhibited a day-to-day variability of 16%, while they ranged from 3500 ng L^{-1} to 4700 ng L^{-1} in STP Dortmund. In the effluent the concentrations were lower, ranging from 840 ng L^{-1} to 1800 ng L^{-1}. From these concentration data, elimination rates were calculated on a daily basis. These daily elimination rates ranged from 61% to 81%; thus, breakthrough rates of 19% to 39% have to be taken into account. The inflow concentrations found in this STP were lower than those determined by Simonich et al. [26] as about 9 µg L^{-1}, but it should be noted that the plants studied by that group operated mainly on domestic wastewater, while the one used for Table 2.14 operated on 50% domestic and 50% brewery wastewater. Thus, the concentrations in this plant are lower in all contaminations than in similar plants. On the other hand, the effluent concentrations were lower than those determined in Europe as well [26].

The concentrations of OTNE in the wastewater inflow as well as in STP effluent were significantly higher than those reported for HHCB or AHTN or the bactericide triclosan in the same samples [38, 68]. The elimination rates were typical for state-of-the-art STPs in this region.

Tentative experiments on sorption of OTNE resulted in sludge concentrations of 2900–4500 ng g^{-1}, which is again higher than those found for HHCB or AHTN. The analysis of mass flows for the whole period in the STP results in an introduction of 3 kg during the four-day period. Thirty-four percent (1.0 kg) of the OTNE introduced to this STP is directly discharged, while about 26% (0.8 kg) of the incoming material is sorbed to the sludge. The residue of the incoming material is not recovered in this study, which may be a function of measurement uncertainties as well as of degradation/transformation effects.

OTNE in Surface Water
The results of the sampling campaign of surface waters and STP effluents are demonstrated in Fig. 2.23. In these samples concentrations from 100 ng L^{-1} to 900 ng L^{-1} were determined. The highest concentrations were determined in the effluents of STPs such as STP Ölbachtal (stations 56–59) and the tributary Neger (station 29) as well as in samples from the tributary Lenne (stations 46/47), which is heavily influenced by STPs. The concentrations in the upper part of the river are below the limit of quantification, while concentrations of 100–250 ng L^{-1} were found in typical surface water samples from the Ruhr after the passage of the first major cities. The concentrations are thus considerably higher than those determined for the polycyclic musk compound HHCB in the same samples [47, 68].

Mass Balance Considerations/per-capita Emissions
In the four-day balancing experiment (Table 2.14), nearly 3 kg OTNE reached the STP, and it released about 1 kg to the receiving waters. This would indicate consumption rates of about 0.86 g per capita annually in the area of Dortmund. The effluent data gained in the balancing experiment compare well with those obtained from the effluents from stations 56–59 in the surface water experiment

(Fig. 2.23). Estimates on mass transports from the Ruhr concentration data result in discharges of about 440 kg from the Ruhr to the Rhine. Considering the elimination rate of about 66% in the STPs, this leads to annual consumption of 0.5 g Iso E Super per capita. On the other hand, discharge rates might be interesting to know. Discharges can be obtained in two ways: (1) by assessing the effluents of STPs or (2) from transports in the rivers.

Considering the effluent of STP Dortmund, each citizen discharges 0.25 g annually to the surface waters. The emissions from STP Bochum were of the same order of magnitude, estimated to be 0.15 g per capita (Fig. 2.23). This plant operates an extensive pond system after the final clarifier, which may increase the elimination rates. On the other hand, a comparison of mass transports in the Ruhr with the number of inhabitants that discharge treated wastewater to this river leads to an in-river transport, and thus a discharge, of about 0.25 g annually per capita. Both numbers were derived independently and their difference is small; therefore, it seems that the emission rates can be considered thus established (0.15–0.25 g/a per person) for the Ruhr drainage basin.

Because Iso E Super (OTNE) is present in most aquatic ecosystems, it will raise new issues for both wildlife and drinking water production. It may be important to search for alternatives that are better eliminated in STPs and less stable in the environment.

2.1.6
Other Fragrances: Nitroaromatic Musks and Macrocyclic Musks

Sensoric effects similar to those of the polycyclic and nitroaromatic musks can be reached with a multitude of fragrance compounds. HHCB and AHTN make up one group of these compounds (polycyclic musks).

In earlier years the nitroaromatic compounds musk xylene and musk ketone (structural formula shown in Fig. 2.24) caused concern. After a multitude of data was presented, musk xylene was voluntarily withdrawn from the German market. Its replacement was so effective that nowadays the inflow concentrations of musk xylene in STPs in Germany are far below those of the polycyclic ones (about 50 ng L^{-1} at STP Dortmund and about 13 ng L^{-1} in the effluents). In sludges a concentration of about 3 ng g^{-1} (dry weight) musk xylene was determined. The concentrations in STP A and STP B (sampled in spring 2003) are in the same range. About 25 ng L^{-1} and 24 ng L^{-1} were determined in the influent, while in the effluent 1 ng L^{-1} and 3 ng L^{-1} were determined, respectively (Fig. 2.25). In these cases elimination takes place in the primary aeration/settlement basin as well as in the main aeration basin. The main elimination process, however, is a reduction to mainly musk xylene-4-amine, which is toxicologically more potent than the parent compound.

Further nitroaromatic musks were found in low concentrations. The concentrations of musk ambrette were $\sim$5 ng L^{-1} and those of musk moskene were <5 ng L^{-1} in the influent.

Macrocyclic musk compounds

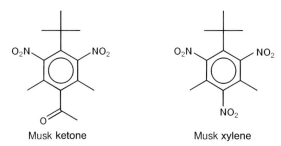

Habanolide   Cyclopentadecanolide

Ethylenbrassylate

Nitroaromatic musks

Musk ketone   Musk xylene

Fig. 2.24 Structural formulas of macrocyclic and nitroaromatic musk compounds.

Table 2.14 Balance of musk xylene and musk ketone in STP Dortmund during a 5-day experiment with the respective day-to-day variability.

	Inflow (g)	Effluent (g)	Sludge (g)	Balance (g)
Musk xylene	44	12	1	−31
Musk xylene-4-amine	0	7	Not analyzed	+7
Musk ketone	63	30	3	−30

The concentrations of musk xylene in surface waters are currently about 1 ng L^{-1} in Germany, while the concentration of its metabolite is of a similar order of magnitude. This was different in the past, when musk xylene-4-amine could be detected in surface waters. For example, in the Elbe estuary, which was supposed to contain diluted river water, the concentrations were around

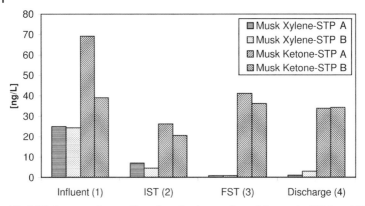

Fig. 2.25 Concentration profiles of musk xylene and musk ketone in STP A and STP B.

0.3 ng L^{-1} for musk xylene, 0.9 ng L^{-1} for musk ketone, and 0.4 ng L^{-1} for musk xylene-4-amine. The concentrations in the inner German Bight in 1990 and 1995 (station 30) were about 0.3 ng L^{-1} musk ketone and 0.3 ng L^{-1} musk xylene-4-amine (unpublished data).

High concentrations will probably still be determined in countries in which musk xylene was not replaced, such as the U.S. and Canada and some European countries. Musk xylene was prioritized under the OSPAR convention as well under the Water Framework Directive.

Musk ketone, on the other hand, is still contained in a multitude of fragrance products. Thus, the concentrations in STP inflow ranges in Dortmund were around 70 ng L^{-1}, while about 33 ng L^{-1} was determined in the effluent. Sludge concentrations were around 9 ng g^{-1} dry weight in 2002. In STP A and STP B, the inflow concentrations were 69 ng L^{-1} and 39 ng L^{-1} in 2003, respectively. The effluent concentrations in both plants were 34 ng L^{-1}. Thus, they were considerably higher than those for musk xylene (but lower than those for polycycles nowadays). The elimination of musk ketone was not very pronounced in STP B, while some elimination did occur in STP A as well as in Dortmund. Elimination of musk ketone takes place partially as sorption to sludge and partially as chemical transformation. A detailed discussion on these STPs is given in Section 2.1.1. These low elimination rates of musk ketone in STPs led to concentrations of 1–2 ng L^{-1} in 2002 in Ruhr River water after passage of the Ruhr megalopolis. The concentrations of musk xylene in the Rhine were slightly lower (0.5–1 ng L^{-1}). Thus, the Rhine transports about 0.07 t annually to the North Sea.

The future of fragrancy with regard to musky compounds probably lies in macrocyclic musk compounds, which in basic functionalities are very close to the original musk obtained from, e.g., the musk deer. These compounds are large, macrocyclic lactones and ketones, such as those marketed under the names habanolide, ethylene brassylate, and cyclopentadecanolide (structural formulas displayed in Fig. 2.24). Comparison of these compounds in OECD-like degradation experiments with activated sludge revealed, indeed, that these macrocycles were eliminated from the respective degradation experiments in closed bottles within hours, while it took several days to eliminate the polycyclic musks. However, it is not yet known which way these compounds are transformed or mineralized. Especially for the macrocyclic compounds, a ring opening and transformation to the respective acids might be assumed. Such a reaction would deliver fatty acid compounds that would be hard to trace through the STP process, as multitudes of fatty acids from other origins are present.

However, until now it was not possible to detect macrocyclic musk fragrances in either sewage treatment plants or surface waters, though the limits of quantification were rather low (i.e., habanolide: 10 ng L^{-1}; cyclopentadecanolide: 3 ng L^{-1}; ethylene brassylate: 10 ng L^{-1}). The market share of these compounds is probably currently too low to be determined in a way similar to the polycycles and nitroaromatic musks.

2.1.7
Behavior of Polycyclic and Other Musk Fragrances in the Environment

Polycyclic musk fragrances such as HHCB and AHTN are used in large dosages in industrial countries. At least in Germany the polycycles have replaced the nitroaromates musk xylene and musk ketone quantitatively. The next probable step in the succession of fragrance products (towards OTNE) is currently taking place in the European market. The step away from the nitromusks can be considered an improvement of the environmental situation because nitro musks, especially musk xylene, may contribute to elevated tumor proliferation rates [72].

On the other hand, the high loads of polycycles and their transformation products such as HHCB-lactone are persistent in the aquatic environment and can reach remote ecosystems such as the North Sea (see Section 2.4.6). In some STPs this lactone can be generated from HHCB in a process that is enantioselective; thus, a biological transformation is highly probable. Extensive studies show that the most important process in STPs considering the fate of polycyclic musk fragrances is sorption to sludge, while transformation (degradation) contributes about 5–10% to the whole balance.

In-river elimination was not determined at a significant level in the studied ecosystem. Another fragrance compound that may be considered polycyclic, i.e., OTNE marketed as Iso-E-Super, has been detected in environmental samples. It must therefore be assumed that all polycyclic musk fragrances are potential contaminants of surface waters as well as sludges and sludge-amended soils. In batch experiments the elimination of macrocyclic musk fragrances (habanolide, ethylene brassylate, cyclopentadecanolide) was much more pronounced than those for the polycycles or nitroaromates. Thus, with regard to elimination in sewage treatment processes, macrocyclic compounds may be a better choice than polycyclic ones.

Concerning transports to the German Bight of the North Sea, it should be noted that the Rhine alone transports about 0.5–2 t/a of each of the polycycles to the North Sea. Considering the residence time of about one year in the German Bight and the water volume contained in it (about 2×10^{12} m^3), this would lead to concentrations of about 1 ng L^{-1} of the respective compounds. On the other hand, only ~0.1 ng L^{-1} was determined, which leads to the conclusion that some elimination mechanisms for the polycyclic musk fragrances from the water phase exist. Half-lives would be on the scale of several months to one year, though.

2.2
The Bactericide Triclosan and Its Transformation Product Methyl Triclosan in the Aquatic Environment

2.2.1
Bactericides from Personal Care Products in Sewage Treatment Plants

Triclosan (2,4,4'-trichloro, 2'-hydroxy-phenylether) (structural formula in Fig. 2.26) is currently used as a bactericide in a multitude of consumer products such as toothpaste, detergents, plastic cutting boards, and sports clothing such as underwear, socks, and shoes [73]. It has been detected in STPs in various countries such as the U.S. [74, 75], Sweden [27], and Switzerland [76]. Additionally, it has been detected in surface waters of the U.S. [113] as well as in Switzerland [76]. Hale and Smith [77] found this bactericide in biota (fish) samples in the Chesapeake Bay.

Concern has been rising in the last few years because this bactericide was detected not only in wildlife near STPs but also in human milk [73]. Additionally, it has been found that this compound has adverse effects on a diversity of organisms at relevant concentrations [78]. Thus, some effort was taken to measure elimination of triclosan from wastewater with the ^{14}C-labeled compound in controlled laboratory experiments [79]. However the concentrations in these experiments were some orders of magnitude higher than in environmental samples.

In sewage treatment plants triclosan may be methylated by metabolic processes. The reported concentrations of triclosan in sludge are around 10 μg g^{-1} in the U.S. [75].

Systematic data on balances of triclosan in an STP and a comparison of diverse sludges, as disposed from the STPs, are missing so far. This study aims to fill this gap with a five-day balance on a large STP operating on 350 000 inha-

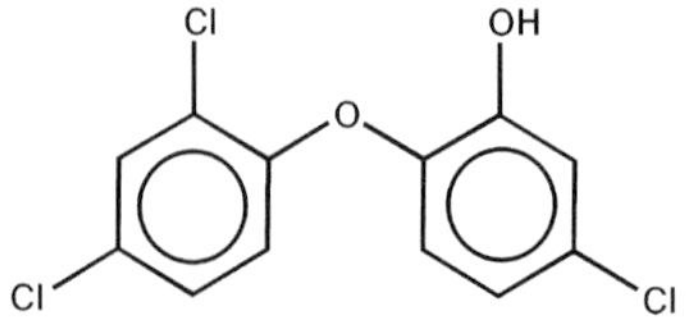

Fig. 2.26 Structural formula of triclosan.

bitants' wastewater. Additionally, the sludge of 20 STPs has been analyzed to obtain comparison data on plant-to-plant variability.

2.2.1.1 Materials and Methods

Samples for the balancing experiment were taken at an STP processing 200 000 m^3 wastewater per day that is located in the vicinity of the city of Dortmund, Germany (compare Section 2.1.1). Additionally, sewage sludge samples were taken by the local EPA authorities from 20 different STPs and sent to the institute for analysis. Data on polycyclic musks from theses samples were reported in Section 2.1.2.

The quantification of triclosan was performed by means of GC-MS (Section 2.1.2). The mass spectrometer was operated at 400 V on the photomultiplier, at 200 °C source temperature, 250 °C interface temperature, and 78 ms dwell time on each ion in selected ion monitoring (SIM) mode. In Fig. 2.27 a spectrum of triclosan obtained from a wastewater sample is displayed, while in Fig. 2.28 a sample chromatogram of a sludge sample extract is shown in comparison to a chromatogram of a standard solution.

The calibrations were performed as a multi-step internal standard calibration. The whole extraction and cleanup procedure was validated with pure water and gave recoveries of 88%. Full quality data of the method obtained from three re-

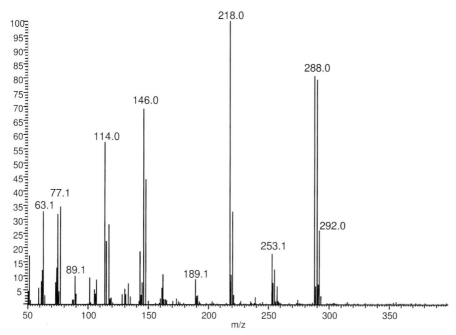

Fig. 2.27 Mass spectrum of triclosan obtained from a wastewater sample. (Reprint with permission from [38]).

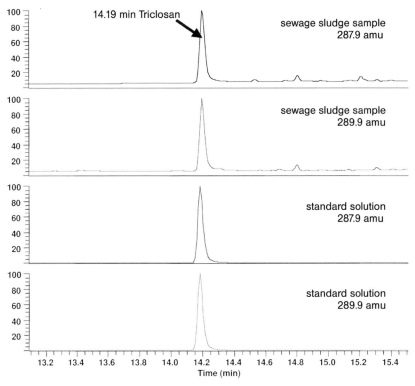

Fig. 2.28 Mass fragment chromatogram (measured in SIM mode) of triclosan from a sludge extract and a pure standard solution. (Reprint with permission from [38]).

Table 2.15 Method validation data for the extraction of triclosan from water and sludge.

Medium	Analyzer mass (amu)	Verifier mass (amu)	Recovery rate (%)	SD	RSD	LOQ
Water	287.9	289.9	88	11	12	3 ng L^{-1}
Sludge	287.9	289.9	94	25	27	4 ng g^{-1}

plica extractions at five different concentrations are given in Table 2.15. The limit of quantification (LOQ) was obtained from the calibration as the lowest concentration giving a signal-to-noise (s/n) ratio ≥10. The spiking levels were 10–10 000 ng L^{-1}. The method was also tested for influence of changes in pH (5–9), humic compounds, and detergent concentrations (data not shown). The recovery rate did not change under these conditions. No significant effects of humic compound and surfactant concentrations on recovery rate occurred for the liquid–liquid extraction used.

The sludge samples were obtained from the loading of trucks. This was the final product after filtering, still containing about 60% water. The extraction and cleanup procedure is described in detail in Sections 2.1.1 and 3.3. The samples were extracted by means of a Soxhlet, and a cleanup was performed by using size-exclusion chromatography coupled with silica.

The recovery rates obtained from three replicate extractions at five different concentrations were 94%. Full method validation data are shown in Table 2.15. The recovery rates were obtained by spiking manure/soil mixtures and re-extracting the respective samples. Spiking levels reached 4–1000 ng g^{-1}.

The pure standard compound and the internal standard (D$_{15}$ musk xylene) were obtained from Ehrenstorfer (Augsburg, Germany). Methyl triclosan was synthesized by methylation of triclosan with trimethylsulfonium hydroxide. The reaction was complete, and the analyte could easily be detected by GC-MS.

Ethyl acetate and cyclohexane were used in analytical-grade (p.a.) quality, while toluene and *n*-hexane were used in residue-grade (z.R.) quality. All solvents were purchased from Merck (Darmstadt, Germany).

2.2.1.2 Triclosan Balances in a Sewage Treatment Plant

During the balancing experiment, the inflow concentrations ranged around 1200 ng L^{-1} (Table 2.16), while the concentrations in the effluent were about 50 ng L^{-1}. The removal rate calculated from these data was about 96%, while a "breakthrough" was determined to be about 4%, respectively. These data are similar to those published recently for a small Swiss STP [80]. On the other hand, the concentrations in the sludge were about 1200 ng g^{-1} dry weight. The concentrations in the sludge as well as in the water samples exhibited very little variation, as the sampling period was a very dry period with no rain at all. Thus, the flow of water and the reactions in the digesters was constant.

The basic data of this STP are documented in Section 2.1.1. With these data, total balances, to estimate the major pathways of triclosan on the STP, were cal-

Table 2.16 Concentrations of triclosan in wastewater and sewage sludge during the balancing experiment. The sewage sludge data refer to dry weight.

Date	Wastewater inflow (ng L^{-1})	Wastewater effluent (ng L^{-1})	Elimination (%)	Breakthrough (%)	Sewage sludge (ng g^{-1})
08. 04. 02	1300	59	95	5	1200
09. 04. 02	1200	58	95	5	1200
10. 04. 02	1100	43	96	4	1000
11. 04. 02	1200	50	96	4	1300
12. 04. 02	1100	43	96	4	1200
Mean	**1200**	**51**	**96**	**4**	**1200**
SD (day-to-day-variation)	80	7.7	0.5	0.4	130

Table 2.17 Balances of triclosan in the STP.

	Inflow (g)	Outflow (g)	Sludge (g)	Sorption to sludge (%)	Balance (g)	Balance (%)
08.04.02	240	11	69	29	−160	−66
09.04.02	230	11	49	22	−170	−74
10.04.02	200	8	67	33	−130	−63
11.04.02	210	9	91	43	−110	−53
12.04.02	200	8	47	25	−140	−72
Sum (5 days)	**1100**	**47**	**330**	**30**	**−710**	**−65**

culated (Table 2.17). All values are rounded at 10%. The balances are calculated on a daily basis and are summed up for the five-day period. In the final balance, negative contributions indicate either losses from chemical conversion of the respective compound or unaccounted losses, e.g., to the atmosphere. During the experiment about 1100 g triclosan entered the STP together with the wastewater. About 50 g of this bactericide was emitted from the plant to the river. About 300 g triclosan was transported with the sludge to final disposal facilities. On the other hand, 65% of the inflowing material "disappeared," i.e., it was found neither in the sludge nor in the effluent water. It is highly probable that the triclosan was transformed (metabolized) and emitted in other forms.

It has been reported in the literature that triclosan may be methylated during the digesting process, and thus the respective methyl ether may be formed [76]. This compound was found in low concentrations in the effluents; however, this transformation product contributed about <0.1–1% of the triclosan balance in this STP with regard to the inflow immissions. Possibly, some bound residues of triclosan were formed in this plant.

This hypothesis was studied by performing a hydrolysis step with the sludge samples that were already extracted with ethyl acetate. This hydrolysis was performed by refluxing the samples for 12 h with concentrated sodium hydroxide in methanol. A similar step was performed for the first (waste) fraction of the size-exclusion chromatography to test for triclosan residues bound to extractable large molecules. About 5% of the amount of triclosan found in the sludge was released by these experiments. Thus, only small amounts of triclosan are transformed to hydrolyzable bound residues in STPs. Whether or not other bound residues may be formed needs to be studied in the future.

These results mean that about 50% of the triclosan is transformed into unknown metabolites or strongly bound residues. These may be dehalogenated products, such as less-chlorinated biphenyl ethers, that may occur under the anaerobic conditions of the digester, or possibly the aromatic ring system is cleaved during the aerobic part of the STP. Whether or not these compounds are soluble and will be found in the aqueous phase will be a topic of further studies. Similar results have recently been published by Singer et al. [80] for Swiss STPs. The emission per capita in this study is slightly lower than those in the Swiss study.

2.2.1.3 **Triclosan in Multi-step Processes in Sewage Treatment Plants**
To study by which means and mechanisms triclosan may be eliminated from the wastewater stream, another study was undertaken with two similar STPs differing in two ways:

1. STP A processed wastewater with a two-step biologically activated sludge treatment to eliminate easily degradable carbon more quickly, while Dortmund STP and STP B had only one activated sludge step.
2. STP A and STP B had biofilters as a final cleaning step (see Section 2.1.1).

Water from those two STPs from diverse stages of the process was sampled (compare Fig. 2.4). Both STPs are operated in the Rhine-Ruhr region and discharge into the Rhine. Both are technologically current state activated sludge plants, i.e., two steps vs. one step biological treatment, but the plants differ in one essential point of the design.

It was thus possible to determine whether biological processes or pure-phase separation was the most effective way to eliminate triclosan from water. Both plants operate on a mixture of industrial and household wastewaters. Thus, the respective inhabitant equivalents equal 1 100 000 and 1 090 000 for STP A and STP B, respectively. Both plants are described in more depth in Section 2.1.1. The same methods for quantification as in Section 2.2.1 have been applied.

STP A (Two-stage Biological Process)
The concentrations in the inflow of STP A ranged around 7300 ng L^{-1} (Fig. 2.29) with little (1500 ng L^{-1}) day-to-day variation. The concentrations were thus considerably higher than those determined in STP Dortmund (Section 2.2.1), but the variation was similar. A sharp decline in concentration (3200 ± 1000 ng L^{-1}) was observed after the first aeration basin. The main aeration basin was even more effective, as the concentrations of triclosan were reduced to 400 ± 100 ng L^{-1}. The fi-

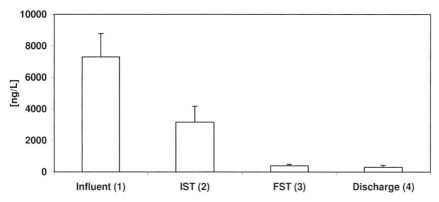

Fig. 2.29 Concentrations (ng L^{-1}) of triclosan in diverse steps of the sewage treatment process in STP A. Error bars indicate day-to-day variations. (Reprint with permission from [81]).

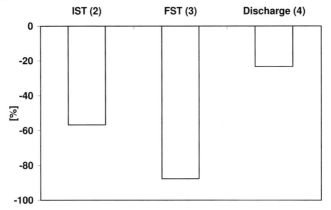

Fig. 2.30 Elimination (%) of triclosan in diverse steps of the sewage treatment process STP A. (Reprint with permission from [81]).

nal filter was not very effective in further reducing the concentrations of triclosan (300 ± 100 ng L^{-1}). Variations in this dataset are all day-to-day variations. In Fig. 2.30 the respective partial elimination rates are shown. Each removal rate refers to the effluent of the previous step in the STP. High removal efficiency was demonstrated in the main aeration basin, but the other parts of the plant also contributed to the removal. The overall removal rate was >95%.

STP B (One-stage Biological Process)
The situation in STP B was similar, thought the concentrations were somewhat lower (Fig. 2.31). The inflow concentrations ranged around 4800 (± 550) ng L^{-1}. On the other hand, the concentration after the primary sedimentation tank (no biological treatment) was relatively high, i.e., 3300 (± 950) ng L^{-1}. The main aeration basin was very effective in removing triclosan; the concentrations after

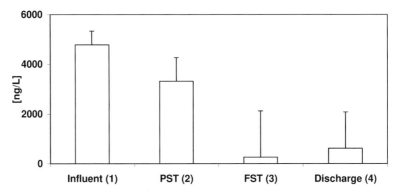

Fig. 2.31 Concentrations (ng L^{-1}) of triclosan in diverse steps of the sewage treatment process in STP B. Error bars indicate day-to-day variations. (Reprint with permission from [81]).

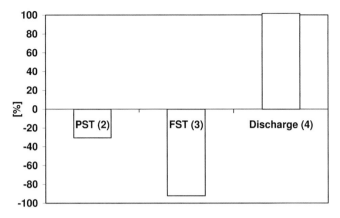

Fig. 2.32 Elimination rates (%) of triclosan in diverse steps of the sewage treatment process in STP B. (Reprint with permission from [81]).

this step were 260 (±1900) ng L^{-1}. After the final filter the concentrations reach 620 (±1500 ng L^{-1}).

These findings are also displayed in Fig. 2.32, in which the elimination rates for the respective steps in the STP are demonstrated. The relatively high emission rate of the final filter should be compared to the high day-to-day variation in this part of the process and is thus not significant. As earlier experiments have demonstrated, bound residues of triclosan can be formed during the passage of the STP, and these bound residues can be released under changing environmental conditions (Section 2.2.1).

In both STPs the total elimination of triclosan was >90% and thus was similar to the plant in Dortmund, though the concentrations as well as the apparent usage per capita were higher. Considering the concentrations that reach the sewage systems, it may be concluded that the per capita annual release rate of triclosan is 1 g in STP A, 0.75 g in STP B, and 0.3 g in Dortmund.

Possibly the usage rate of triclosan reflects socioeconomic data, as the average income in STP A and STP B is higher than those in Dortmund, with its high unemployment rate. However, it is also possible that some degradation in the sewer system may happen and may be different for the respective cities.

However, it should be noted that during the elimination of triclosan, methyl triclosan is formed (Figs. 2.33 and 2.34). Though the concentrations of this transformation product are low, this compound can be easily detected in fish samples, which is a good indication for a persistent and bioaccumulating compound. It might be interesting to note that this compound is formed in the main aeration basin, which might be puzzling because methanogenic, i.e., anaerobic, conditions generally are needed for methylation processes. However, both plants operate simultaneous nitrogen removal in the main treatment basin, which is thus essentially only partially aerobic while it is anaerobic in other parts.

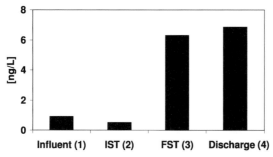

Fig. 2.33 Concentrations and generation of methyl triclosan in STP A.

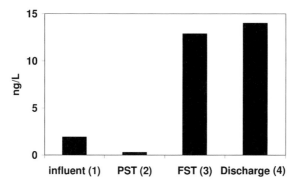

Fig. 2.34 Concentrations and generation of methyl triclosan in STP B.

2.2.2
Triclosan in Sewage Sludge

Sewage sludge samples of 20 STPs of different sizes (10 000 to 1 000 000 inhabitant equivalents) and operating principles (samples from various points in the respective plants; hygienized sludge [calcium carbonate] and non-hygienized sludge; after anaerobic stabilization and without anaerobic stabilization) were taken randomly from a larger set of samples and analyzed to determine whether the concentrations found in the STP used for balancing were representative for other STPs. The concentrations found in these 20 samples ranged from 400 to 8800 ng g^{-1}. Thus, the plant under observation is probably representative for a multitude of plants, although the concentrations in this plant were at the low end of the control group. Some samples exhibited the same concentrations as those published for the U.S. (10 μg g^{-1}), while several others are slightly lower [75]. The distribution of concentrations is shown in Fig. 2.35. It may be worth noticing that the concentration pattern of triclosan was similar to those obtained for polycyclic musk compounds in the same samples, though the concentrations of triclosan are lower. This may be taken as a strong indication that both compounds stem from similar sources, probably domestic usages.

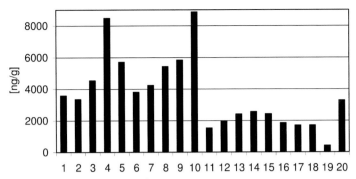

Fig. 2.35 Concentrations (ng g^{-1}) of triclosan in 20 different STPs in North Rhine-Westphalia. Each data point was generated from a double extraction. (Reprint with permission from [38]).

The lowest concentration originates (station 19) from a very small plant in a rather rural area, while the samples with the highest concentrations originated from huge plants in urban sites with 50 000 to 190 000 inhabitants.

2.2.3
Triclosan in Surface Waters

In this part of the study the fate of triclosan and its known transformation product methyl triclosan was studied in surface waters and discharges to the river. A characterization of sampling sites is given in Section 2.1.3, in Fig. 2.11, and in Table 2.7. The same extraction parameters as in previous studies have been utilized (Section 2.2.1), although some experimental parameters were different in comparison to the older studies. Final analysis was performed by quadrupole GC-MS (Thermo-Finnigan DSQ) equipped with a PTV injector and an AS 2000 autosampler. The PTV (4-µL injection volume) was operated in PTV splitless mode applying the following temperature program: 90 °C (0.1 s) → 12 °C s^{-1} → 280 °C (1 min) (cleaning phase; open valve). The GC separation was performed utilizing a DB-5MS column (J&W Scientific, Folsom, CA, USA) (length: 15 m; i.d.: 0.25 mm; film: 0.25 µm) and a temperature program of 100 °C (1 min) → 30 °C min^{-1} → 130 °C → 8 °C min^{-1} → 220 °C → 30 °C min^{-1} → 280 °C with 1 mL min^{-1} He as carrier gas. The transfer line was held at 250 °C, which is sufficient to transfer all compounds from the gas chromatograph into the MS as the vacuum builds up in the transfer line. The ion source was operated at 250 °C. For quantification, the mass fragments 288.0 and 290.0 amu were analyzed for triclosan and mass fragments 302.0 and 304.0 amu were analyzed for methyl triclosan. All fragments were determined in selected ion monitoring mode (SIM) with 60 ms dwell time and the multiplier set to 1089 V (auto tune). Full quality data of the method obtained from three replica extractions at five different concentrations are given in Table 2.18.

Table 2.18 Quality assurance data for the analysis of triclosan and methyl triclosan.

	Analytical ion (amu)	Retention time (min)	Recovery rate (%)	SD (%)	LOQ (ng L^{-1})
Triclosan	288	11.17	88	11	3
	290	11.17	88	11	3
Methyl triclosan	302	11.41	102	11	0.3
	304	11.41	102	11	0.3

Triclosan was identified by its respective mass spectral data and its retention time (because its mass spectral data have already been discussed in earlier sections as well as in the literature [38], these data are not shown here). The mass spectrum of methyl triclosan is shown in Fig. 2.37, while in Fig. 2.36 comparison of retention times of GC peaks obtained from an STP effluent and a standard solution is shown.

The concentrations in the Ruhr, which provides several million inhabitants with drinking water though about 2 million inhabitants discharge treated waste water to this river are displayed in Fig. 2.38 together with some STP effluents. The concentrations of triclosan ranged from below 3 ng L^{-1} to 10 ng L^{-1} in true surface water, while values up to 70 ng L^{-1} were found in STP effluents such as Bochum-Ölbachtal. Thus, the effluent concentrations of the STP studied in Section 2.2.1 (see Table 2.17) are typical for this region. High concentrations were also detected for the tributary Lenne, which is heavily influenced by STP effluents at this location. These discharge data are lower than those determined for STP A and STP B, which were 300 ng L^{-1} and 600 ng L^{-1}, respectively. The concentrations of methyl triclosan ranged from <0.3 ng L^{-1} to 5 ng L^{-1} in true surface water samples, while in effluent samples concentrations up to 20 ng L^{-1} were determined.

The values of methyl triclosan were about 40% of the triclosan concentrations in the effluents as well as in most surface water samples. Some variability was determined in the emission patterns of the respective STPs. It should be noted that neither the concentrations nor the pattern changed drastically after the immissions of the largest STP on the river (stations 57–59), but the concentrations stay constant at about 5 ng L^{-1} triclosan and 2 ng L^{-1} methyl triclosan.

Fig. 2.36 (a) Comparison of mass fragment chromatograms (SIM mode) of methyl triclosan (Triclosan-Me) in wastewater extract and sample solution.
(b) Comparison of mass fragment chromatograms (SIM mode) of triclosan from wastewater sample extract and standard solution.
(Reprint with permission from [81]).

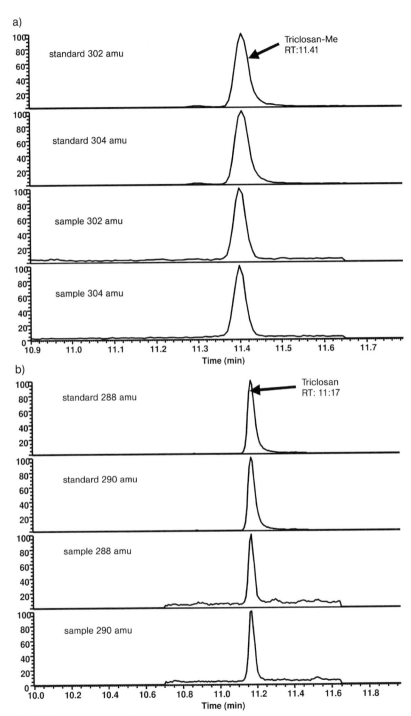

Fig. 2.36 (legend see p. 64)

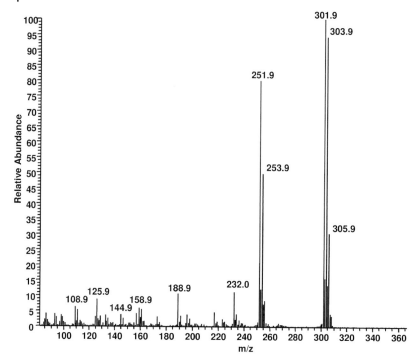

Fig. 2.37 Mass spectrum of methyl triclosan. (Reprint with permission from [81]).

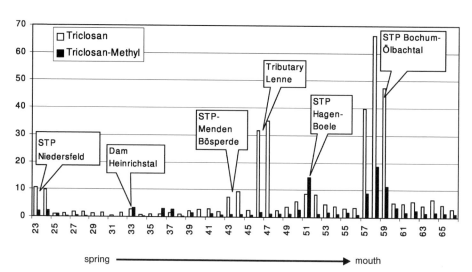

Fig. 2.38 Concentrations (ng L^{-1}) of triclosan and methyl triclosan in Ruhr water, in STP effluents (indicated as STP), and in tributaries and lakes. Characteristics of the sampling stations are given in Table 2.7 and Fig. 2.11. (Reprinted with permission from [81]).

2.2.3.1 Estimation of Elimination Constants for Triclosan in a River

All the lakes that the Ruhr passes are shallow, artificial lakes with eutrophic characteristics that were constructed in the first half of the last century. These ecosystems thus differ considerably from those studied in Switzerland, e.g., the Glatt River and the Swiss lakes [76, 80]. However, for the water in the river dataset presented above, degradation kinetics can be assessed if two points of the river are chosen between which neither clean water input (by tributaries) nor wastewater discharges occur and if the runtime between these two points is known. Thus, two or more data points of the respective time series are generated at the same time, taking the runtime of the river as the time axis. The Ruhr needs 140 h (6 d) [82] from station 60 to station 66. From the triclosan and methyl triclosan data between stations 60 and 66 and the respective runtime, an elimination rate constant k for in-river removal can be calculated as follows, assuming first-order kinetics:

$$k = \frac{\ln\left[\dfrac{C_0}{C}\right]}{t} \tag{1}$$

The respective k values are 0.062 (d^{-1}) for triclosan and 0.066 (d^{-1}) for methyl triclosan.

With Eq. (2) a half-life can be assessed:

$$T_{1/2} = \frac{\ln\left[\dfrac{C_0}{\dfrac{C_0}{2}}\right]}{k} = \frac{\ln 2}{k} \tag{2}$$

The half-life for triclosan as well as for methyl triclosan is about 11 d in this river. This is longer than the runtime of the water (6 d) in the respective area. Thus, the elimination is not apparent in the respective dataset (Fig. 2.37). It seems that in this part of the river elimination processes of both compounds do not control the concentrations in the rivers as much as the immissions do.

These data differ considerably from those determined for completely different ecosystems in Switzerland, in which photolysis of triclosan but not of methyl triclosan is discussed [76, 80].

Also, these results differ from those of creeks such as the Cibolo Creek (Texas, USA) which contains about 100% wastewater [3 million m^3/a]) or the Itter Creek (North Rhine-Westphalia, Germany with 35 million m^3/a containing about 25 million m^3/a wastewater), in which fast elimination (78% within 8 km) has been observed [83]. These creeks held starting concentrations of 400 ng L^{-1} and 200 ng L^{-1}, respectively [83, 84]. These concentrations were 1 to 2 orders of magnitude higher than those determined in the study for the Ruhr River. They are similar to 100% wastewater effluents of highly charged wastewater treatment plants as determined in Section 2.2.1. These concentrations are even higher than any discharge into the Ruhr. Maybe this difference explains some of the differences in elimination behavior.

Also, the discharge or immissions to the river concentration may vary, as shown by the differences between STP A and STP B in comparison to the immissions to the Ruhr. This may be due to differing usage rates in the different areas, as already discussed concerning the musk fragran as in Section 2.1 STP A and STP B in comparison to STP C (Dortmund). But it is also possible that different technologies have some effect on these findings as well. For instance, in Bochum-Ölbachtal the effluent passes a set of multiple final clarifier ponds before the water is discharged into the river. These final ponds may reduce the concentrations of triclosan and methyl triclosan even further than the conventional STP process by activated sludge in aeration basins or trickling filters.

The concentrations found in the Ruhr system are of the same order of magnitude as determined by Lindström et al. [76] for the lakes Greifensee and Zürichsee. Other groups have also studied the fate of triclosan in surface waters, but they have relied on semipermeable membrane devices (SPMD), which are good indicators of bioaccumulation but give less reliable results for direct concentrations [85, 87].

On the other hand, some variability in the patterns of triclosan and methyl triclosan was determined in the Ruhr River ecosystem, but no significant changes were detected during the passage of about 100 km of the river with 8 d residence time, while there were changes in patterns in the Greifensee [87]. This may be attributed to the fact that hydraulic residence times in the Ruhr and its lakes are short in comparison to that of the Greifensee (one year). Another difference is that the lakes in the Ruhr system are rather eutrophic, while the Swiss lakes are oligotrophic.

In the Swiss lakes triclosan and its metabolite are eliminated slowly but with similar rate constants. The rate constants are in the same range as discussed as photolyses rates for triclosan by Tixier et al. [86] for this area. However, according to this work, triclosan should exhibit significantly different elimination rates than its transformation product, which is not the case in the Ruhr. On the other hand, this group also suggested that sedimentation is relevant, which is probably even more likely in the eutrophic situation at the Ruhr River.

Considering these data, it seems that the concentrations of triclosan and its metabolite in the rivers are lower than that of flame-retardants but similar to that of the polycyclic musks. The fate of this bactericide, however, seems to be dependent on the ecosystem and the concentration studied. The concentrations found in the Ruhr may have some impact on fish, which are known to accumulate both triclosan and methyl triclosan [73, 87].

Generally, it seems important to study the fate of triclosan and its transformation product separately in future studies, as at least in ecosystems such as the Ruhr, the two compounds are similarly eliminated, while in others significant differences have been observed.

2.2.4
Discussion on Triclosan and Methyl Triclosan in the Environment

Triclosan is emitted at a rate of 0.25–1 g per inhabitant annually to the waste-water in the study region. These emission rates exhibit much higher variability than those determined for the polycyclic musks (see Table 2.19). This may be due to different usage rates caused by differing socioeconomic data, or it could be due to elimination processes in the sewer system (wastewater pipe system), which might be different in the diverse locations as well.

Elimination rates of triclosan in the STPs are 87–95%, which may be considered good in comparison to other compounds. Thirty percent of the elimination is due to sorption to sludge, while at least 10% is chemically bound to sludge as bound residues. About 1% of the triclosan introduced to the STPs is transformed to methyl triclosan and is found in the discharge of the STPs.

Because methyl triclosan may be bioaccumulated even more than triclosan, the emissions of these compounds are important with regard to contamination of fish. A completely open question is the fate of triclosan in sludge when it is used as fertilizer and added to agricultural land. It may well be assumed that residues of triclosan are persistent in soils and add to the load of pollutants in food. However, the largest source of exposition of the population is the use of triclosan-containing products, as these are generally applied orally or worn on the skin.

Table 2.19 Per-capita emission data for three different STPs in the Rhine-Ruhr region. The fragrance HHCB is given as well for a better understanding.

		Annual usage per person (g)
STP A	HHCB	0.66
	Triclosan	1.04
STP B	HHCB	0.37
	Triclosan	0.70
STP C	HHCB	0.42
	Triclosan	0.25

2.3
UV Filters/Sunscreens

Organic UV-filters such as 4-methylbenzylidene-camphor (4-MBC), ethylhexyl methoxycinnamate (EHMC), octylcrylene (OC), and benzophenone 3 (BP-3) (see Fig. 2.39) are used in sunscreen creams to protect the skin against burns and possibly skin cancer [88]. They are also used in a multitude of cosmetics for various reasons including increased shelf life of the respective product. Balmer

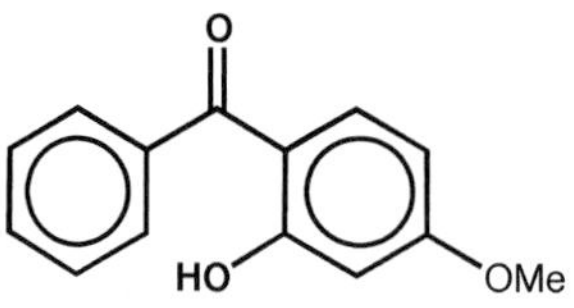

benzophenone-3

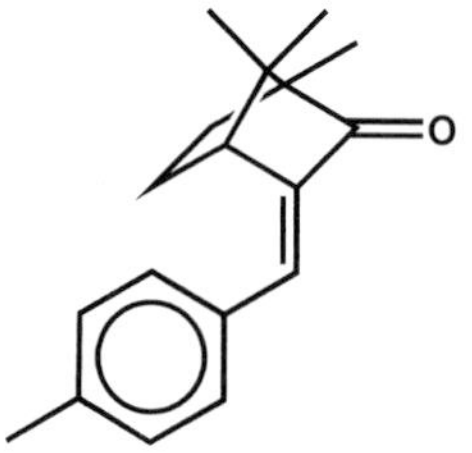

4-methylbezylidene-camphor

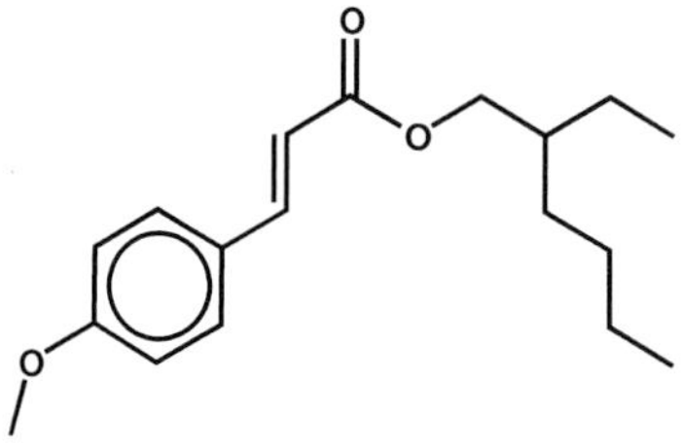

ethylhexyl methoxy cinnamate

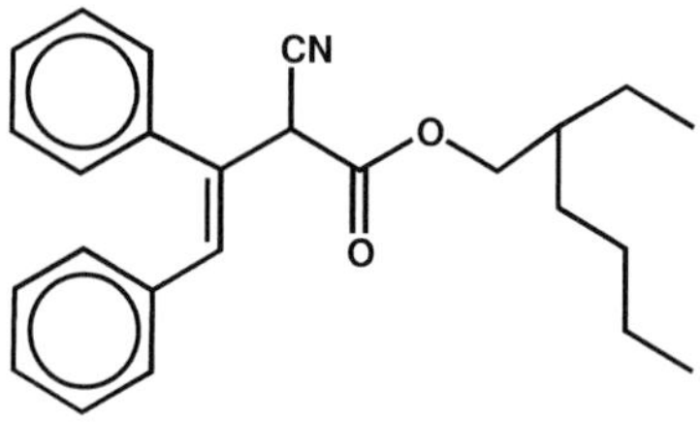

octocrylene

Fig. 2.39 Structural formulae of UV filters.

et al. [88] estimated a rate of 0.5–2.0 g UV filter per application of sunscreen. Obviously, these compounds may enter the aquatic environment via two routes:

1. sunscreen-direct inputs to lakes and rivers during bathing activities on sunny days, and
2. sunscreen, cosmetic products, and showering/washing activities as inputs to wastewater and introduction via STPs into rivers.

These compounds are lipophilic, with log K_{ow} ranging from 3.8 (BP-3) to 6.9 (OC), while 4-MBC and EHMC have log K_{ow} of 5.1 and 6.0, respectively [88]. On the other hand, some of these UV filters have been found by Schlumpf et al. [89] to exhibit estrogen hormonal activity, and thus they are worth consideration.

2.3.1
Endocrine Properties of UV Filters

Schlumpf et al. [89] tested various compounds for hormonal, especially estrogenic, activity. In this study the following endpoints were used:

- MCF-7 cells. This is a human breast cancer cell line that has been used in a multitude of other studies and thus has gained the status of a standard test. Generally speaking, this assay tests for cell proliferation and pS2 protein exprimation.
- E-SCREEN. This is a modified test with MCF-7 cells.

- Uterotrophic assay. This is an assay in which female baby rats were administered feed with the respective chemical. After four days the rats were sacrificed and the uterus removed. The values are compared to negative (no chemical administered) and positive (ethynyl estradiol) controls.

The following UV filters were tested: benzophenone-3 (BP-3), homosalate (HMS), 4-methyl-benzylidine camphor (4-MBC), octyl-methoxycinnamate (OMC), octyl-dimethyl-PABA (OD-PABA), and butyl-methoxydibenzoylmethane (B-MDM). Some of the results are shown in Table 2.20.

In the in vitro cell tests, all UV filters (BP-3, 4-MBC, OMC, OD-PABA, and HMS) except B-MDM were positive, with an effective concentration for 50% of the tests (EC_{50}) of 3.7 µM, 3.0 µM, 2.4 µM, 2.6 µM, and 1.6 µM, respectively. In comparison, 17β-estradiol gave an EC_{50} of 1.2 pM as a control. 4-MBC, HMS, and BP-3 induced proliferation of the pS2 protein considerably.

The in vivo test gave similar findings: 4-MBC, OMC, and BP-3 gave clear positive (i.e., increased uterine weight) results. Effective dose for 50% of the tests (ED_{50}) values of 309 mg kg^{-1} d^{-1}, 934 mg kg^{-1} d^{-1}, and 1250 mg kg^{-1} d^{-1} were determined, with ethynyl estradiol as positive control resulting in 0.82 µg kg^{-1} d^{-1}.

The author also performed dermal expositions and found uterotropic effects in rat at 4-MBC concentrations that are allowed in sunscreen creams. There is

Table 2.20 Estrogenic effects of UV filters (adapted from Ref. [89]).

Compound	In vitro tests				In vivo tests		
	PE [a]	RPE [b]	Maximum cell count increase (% of E_2) [c]	EC_{50}	Increase of uterine weight over control	Maximal weight increase (% of EE_2) [d]	ED_{50} (mg kg^{-1} d^{-1})
17β-Estradiol (E_2)	16.70	100	100	1.22 pM			
17α-Ethinyl-estradiol (EE_2)					4.08	100	0.818 E-3
BP-3	17.49	105.0	95.09	3.73 µM	1.23	7.60	1000-1500
4-MBC	13.49	79.54	87.66	3.02 µM	2.09	35.51	309
OMC	10.72	61.90	77.18	2.37 µM	1.68	22.21	937
OD-PABA	9.13	51.77	55.54	2.63 µM	1.04	1.15	Inactive
HMS	6.78	36.81	79.65	1.56 µM	1.12	3.79	Inactive
B-MDM	4.30	21.01	13.27	Inactive	1.06	2.01	Inactive

a) Proliferation effect over control: PE = (maximum cell count of experimental group/(cell count control).
b) Maximal proliferation effect (% of E_2): RPE = (PE-1)/(PE of estradiol-1) × 100.
c) (cell count of experimental group − cell count control)/cell count estradiol − cell count control) × 100.
d) (uterine weight of experimental group − control)/(uterine weight of ethynyl estradiol − control) × 100.

also evidence of dermal absorption by human skin. The authors concluded that estrogenic activity of UV filters has not yet found its way into legislation on endocrine disruption, especially in Europe [90].

2.3.2
UV Filters in Aquatic Ecosystems

The concentrations of 4-MBC, BP-3, EHMC, and OC in wastewater treatment plant influent and effluent were determined by Balmer et al. [88]. The influents held about 0.6 µg L^{-1} 4-MBC, 0.7 µg L^{-1} BP-3, 0.8 µg L^{-1} EHMC, and 0.2 µg L^{-1} OC in the spring, while the concentrations in other STPs in June were higher than expected (4, 6, 9, and 5 µg L^{-1}, respectively), while the concentrations in late summer (August and September) were lower. They determined the following elimination rates: $72 \pm 27\%$ (4-MBC), $90 \pm 11\%$ (BP-3), $98 \pm 1\%$ (EHMC), and $96 \pm 5\%$ (OC).

The same group also analyzed lake water from the Swiss lakes, which contained mostly 4-MBC and BP-3 with concentrations of 2–35 ng L^{-1}. EHMC and OC were determined at only a few sample stations with concentrations up to 6 ng L^{-1}, but in most samples they were <2 ng L^{-1}.

Nagtegaal et al. [91] were probably the first authors to analyze UV filters in environmental samples. Their work was performed in volcanic lakes in the Eiffel area that are used for recreational purposes. In this work about 2 mg UV filters per kilogram lipids were determined in fish samples. Several fish species were thus contaminated, especially with 4-MBC. These concentrations were in the same range as PCBs and DDT analyzed in the same fish species. This was surprising, as this lake does not receive wastewater. Thus, it may be concluded that recreational and direct introduction of sunscreen to the water was the major method of introduction of these compounds into these lakes.

In addition to the water samples, Balmer et al. [88] also analyzed fish samples from the somewhat larger Swiss lakes. However, only a few numbers are given: 4-MBC: 44–94 ng g^{-1} lipid; BP-3: 66–120 ng g^{-1} lipid; EHMC: 64–72 ng g^{-1} lipid; and OC: 25 ng g^{-1} lipid. The authors concluded that though some of the UV blockers entered the respective lakes via the STPs, there were also some direct inputs. These inputs were estimated to be correlated to bathing and other recreational activities during the summer months.

Balmer et al. [88] assessed a bioconcentration factor (BCF_L) on the basis of the 4-MBC concentrations in fish lipids. The BCF_L was found to be 9700–2300 for roach in Lake Zürich, indicating log BCF_L values of 4.0–4.4. This is considerably lower than expected for such a lipophilic compound (log $K_{ow} = 5.1$). It is also lower than those obtained for methyl triclosan from the same samples (log $BCF_L = 5.0–5.4$) with a similar log K_{ow} (5.0). This would indicate either less accumulation from the water (for unknown reasons) or better elimination (possibly by metabolization) of 4-MBC than of methyl triclosan from the fish. However, it might also indicate de novo formation of methyl triclosan from triclosan in the fish. Balmer et al. [88] compared these accumulation rates to those obtained for semipermeable membrane devices (SPMDs), which are supposed to mimic fish with no metabolization possibilities, as they are polyethylene tubes filled with lipids. In these SPMDs Balmer et al. found much higher concentrations of 4-MBC. These experiments would support the hypothesis of metabolization of 4-MBC in fish.

These data agree with data from the same area published earlier [92]. At that time it was suggested that about 100 kg each of 4-MBC, BP3, OC, EHMC, and OT entered Lake Zürich during the summer of 1998. The situation was similar for the Hüttnersee, which is much smaller: about 1 kg of each of theses compounds were introduced into this lake during the respective season. From these data steady-state concentrations in both lakes were calculated, taking into account the residence time of the water. The predicted concentrations varied from 24 ng L^{-1} for BP-3 in Lake Zürich to 270 ng L^{-1} of EHMC in the Hüttnersee. However, the measured concentrations were considerably lower than the expected values. It was concluded that the UV filters were indeed photolyzed in the waters of the respective lakes. Additionally, it was discussed whether the input estimation might have overestimated the true inputs to the lakes. However, it was noted that the concentrations in the surface layers of the Hüttnersee were much higher than the deep waters in summer. This effect did not occur during the spring and autumn seasons.

2.3.3
Enantioselective Considerations for UV Filters

It might be interesting to note that 4-MBC is a chiral compound with two stereocenters. Buser et al. [93] succeeded in performing the enantioseparation of this compound. While the samples from the respective products were pure racemates, some wastewater treatment plant effluents exhibited slightly changed patterns. However, the comparison of lake samples revealed considerable differences between four lakes. Lake Zürich and the Hüttnersee had few alterations, while the Greifensee showed considerable alterations of the enantiomeric patterns. On the other hand, the Jörisee in the high alpine mountains did not contain any 4-MBC at all, as no wastewater is introduced into this lake and there is no bathing activity. The authors concluded that the Greifensee exhibits the longest residence time of water of all three contaminated lakes and that it is the most eutrophic one. These effects may have led to better biodegradation efficiency than in the other lakes. Also, the authors analyzed some fish samples to gain insight into the enantiomeric composition of 4-MBC in fish. As roach and perch from diverse lakes have been analyzed, no clear picture can be drawn up to now. However, especially the perch samples seem to reveal an enantiomeric ratio of 9:1. In comparison, the wastewater contained racemic 4-MBC with an enantiomeric ratio of 1:1.

2.4

Organophosphate Flame-retardants and Plasticizers

2.4.1

Introduction

2.4.1.1 Flame-retardants

Flame-retardants are used to protect combustible material from burning. They are heavily used for the protection of textiles, construction foams, and all types of polymeric materials, e.g., in cars, furniture, electronics, etc. They make every-day life safer from to fires.

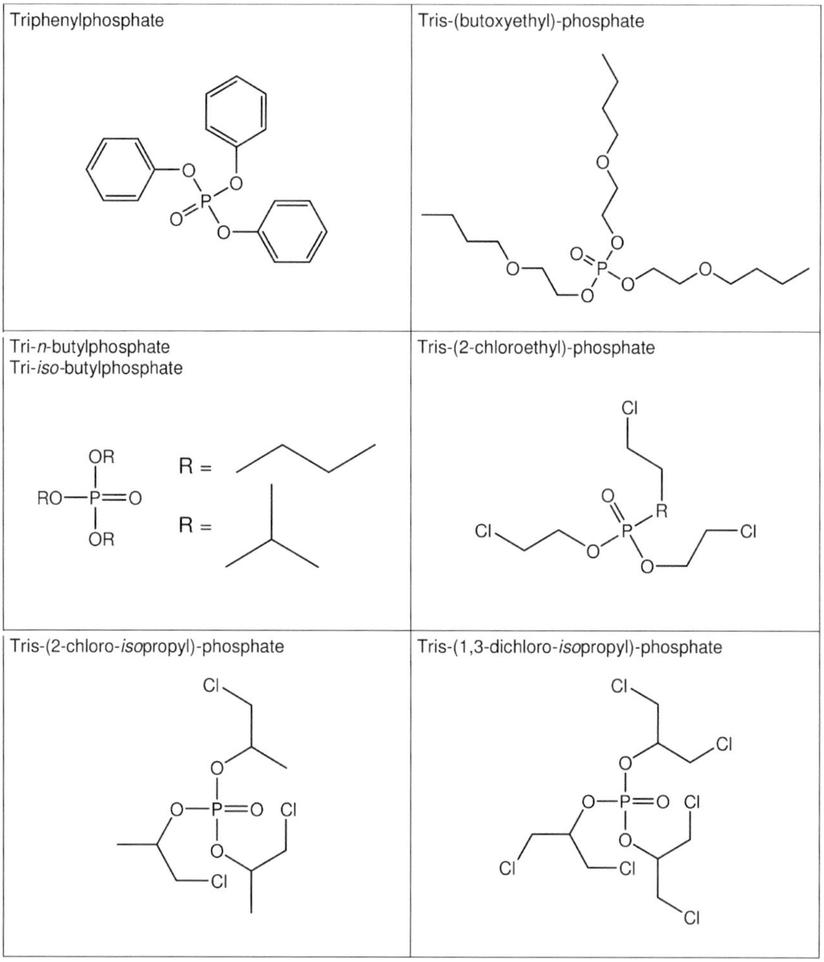

Fig. 2.40 Structural formulae of organophosphates.

There are three main groups of flame-retardants currently on the market: organophosphates, organobromine compounds, and inorganic compounds such as antimony or boron salts. Initially, these compounds were not considered environmentally relevant because they were believed to remain in their respective materials. However, it has turned out that some mobilization does occur. For the aqueous ecosystem the organophosphates (structural formulae in Fig. 2.40) are of special relevance, as they are more water-soluble than the organobromines and are thus assessed to be mobile once they are introduced into aquatic pathways. With regard to the organophosphates, the chlorinated ones are of special concern.

One of the more recently introduced chlorinated organophosphates is tris-(2-chloro-1-methylethyl) phosphate (TCPP), which is used as a flame-retardant agent in flexible and rigid polyurethane foam. It is marketed by various manufactures under brand names such as Fyrol PCF, Antiblaze TMCP, and Levagaard PP. About 7500 t of this product was used in 1998 in Germany [94]. Leisewitz et al. [94] estimated the market of flame-retardants to be 1 million t/a worldwide. Depending on the region, about 14–25% of these is thought to be organophosphates. There have been rapid changes on the market, though, as some years earlier mostly tris-(2-chloroethyl) phosphate was used.

In the whole of Western Europe nearly 23 000 t organophosphates were used in 1998 [95]. Most of the TCPP (up to 95%) [96] is currently used in rigid polyurethane foam plates, which are used as thermal insulation material in construction. Organophosphates such as triphenyl phosphate (TPP) and TCPP have been identified by Carlsson et al. [97], Kemmlein et al. [98], and Marklund et al. [99] in indoor contamination. However, computers and other electronic equipment have been described by the same group to emit predominantly TPP and not TCPP. On the other hand, other not yet fully identified sources may be relevant for the environment as well. The compound has been identified in air pollution–related studies, which were designed for the study of organophosphate pesticides (e.g., Aston et al. [100] determined TCPP in pine needles in the Sierra Nevada).

This compound (TCPP) is described in the literature as only moderately toxic (in fish: NOEC ~ 10; LC_{50} ~ 30 mg L^{-1}), but it is neurotoxic (LD_{50} [oral, rat]: 700–3000 mg kg body weight^{-1}) [108]. The mutagenicity of TCPP is currently under discussion. While in some datasets this organophosphate appeared to be non-mutagenic, in others it exhibited mutagenic properties. The IUCLID datasheet [101] reports alteration of rat testes after administration of TCPP.

Thus, there was high motivation to include TCPP in these studies on STPs [102]. Other chlorinated organophosphates (see Fig. 2.40) such as tris-(2-chloroethyl) phosphate (TCEP) and tris-(1,3-dichloro-isopropyl) phosphate (TDCP) have been used for similar purposes but are being phased out because of discussions on their toxicological properties (see Table 2.21).

Table 2.21 Usage pattern and toxicology of organophosphates.

Acronym/Name	Usage	Toxicology	Remark
TiBP Tri-*iso*-butylphosphate TnBP Tri-*n*-butylphosphate	Lubricant, plasticizer, concrete (pore size regulation) Solvent for cellulose esters, lacquers, and natural gums; primary plasticizer in the manufacture of plastics and vinyl resins; antifoam agent for concrete [104, 105] and hydraulic fluids [106, 107]	Log K_{ow}=4.0 [103] Neurotoxic [107] K_{ow}=4.0 [103]	
TCEP Tris-(2-chloroethyl)-phosphate	Flame-retardant (mostly polyurethane foam) [108]	Carcinogen [108] K_{ow}=1.7 [103]	Phased out due to toxicity issues [108]
TCPP Tris-(2-chloro, 1-methyl ethyl) phosphate; also called Tris-(chloropropyl) phosphate	Flame-retardant (mostly polyurethane foam) [108]	Possible carcinogen [108] K_{ow}=2.6 [108]	Its genotoxicity and carcinogenicity have been debated recently [138]
TDCP Tris-(2-chloro, 1-chloromethyl-ethyl)-phosphate; also called Tris-(dichloro-propyl) phosphate	Flame-retardant (mostly polyurethane foam), textiles, diverse [108]	Carcinogen [108] K_{ow}=3.8 [108]	Used for specialties only
TPP Triphenyl phosphate	Hydraulic fluids and flame-retardant [109, 110]	Possibly neurotoxic [110]) K_{ow}=4.7 [103]	
TBEP Tris-(butoxyethyl) phosphate	Plasticizer (rubber and plastics), floor polish [111]	K_{ow}=3.8 [111]	

2.4.1.2 Organophosphate Plasticizers

The non-derivatized alkyl phosphates such as triphenyl phosphate (TPP), tributyl phosphates (*iso*-and *n*-isomer, TiBP and TnBP, respectively), and tris-(butoxyethyl)-phosphate (TBEP) are predominantly used as plasticizers and lubricants and to regulate pore sizes, e.g., in concrete, though in some cases they are also used as flame-retardants. Figure 2.40 gives an overview of the structural formulae of the organophosphate esters that were analyzed, while Table 2.21 gives the respective usage patterns and information on toxicology.

2.4.2
The Organophosphate Flame-retardant TCPP in a Sewage Treatment Plant

Considering the high production of TCPP, it was decided to study whether TCPP reaches the STPs and, if it does, whether sewage treatment is able to eliminate this compound from the wastewater. For a structural formula, refer to Fig. 2.40.

TCPP has tentatively been identified by several authors in surface waters, but a lack of appropriately pure standards or other problems have resulted in identification but not quantification of this compound in environmental samples [27, 112]. A good set of data is presented by Kolpin et al. [113] on surface water, in which concentrations of < 100–160 ng L^{-1} were reported in the U.S. Similar organophosphates were detected by Fries and Puttmann [114] in Germany at concentrations varying from 17 ng L^{-1} to 1500 ng L^{-1}. These authors gave no clues on the imission pathways, though. Researchers such as Prösch et al. [115] have suggested that these compounds enter the aquatic environment via the STPs, as elevated concentrations were found in the effluent of some plants. In that study it was assumed that the main path of entrance to the sewage system might be from the textile industry. On the other hand, the textile industry is not relevant in the Ruhr area. However, the amount that was released from the plant in this older study remained unclear, as did the degradation capacities of the respective STPs.

After proper identification of this compound from wastewater and sewage sludge extracts, a balance on an estimation of discharged amounts was performed in the current study to acquire new insights into the phase distribution and elimination of TCPP in STPs.

2.4.2.1 Materials and Methods

Samples for the balancing experiment were taken at an STP processing 200 000 m^3 wastewater per day located in the vicinity of Dortmund, Germany. The sampled plant as well as the methods used for determination were described in Section 2.1.1. The plant does not add anything for co-fermentation purposes to the digesters, but it receives wastes from chemical toilets that are introduced to the aeration basins (data obtained from the plant's management). Additionally, sewage sludge samples were taken by the local environmental protection authorities from 20 different STPs and sent to the institute for analysis.

The water samples were immediately extracted with 10 mL toluene after adding an aliquot of internal standard solution (D$_{15}$ musk xylene as well as D$_{27}$ tri-*n*-butyl phosphate). D$_{15}$ musk xylene was chosen as internal standard (IS) because the original method was designed to analyze personal care products. Additionally, D$_{27}$ tri-*n*-butyl phosphate (D$_{27}$T*n*BP) was used as IS. Quantifications with both internal standards gave identical results when recovery rates were taken into account. In the sludge samples, though, D$_{27}$T*n*BP was hard to identify because of overlapping matrix. Thus, D$_{15}$ musk xylene was used as internal standard throughout.

In Fig. 2.41 the mass spectrum of tris-(2-chloro-1-methylethyl) phosphate obtained from an inflow water sample is shown, while in Fig. 2.42 a sample chromatogram of a wastewater inflow sample is shown in comparison to a standard solution.

This procedure was validated with pure water and gave recoveries of 71%. All data were checked successively for compliance with the isotopic distribution of 277 versus 279. Full quality data of the method obtained from three replicate ex-

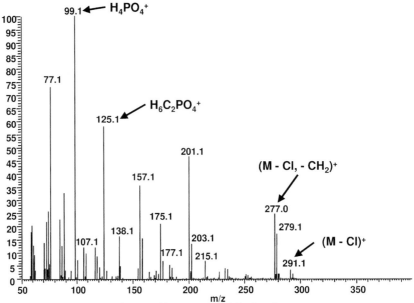

Fig. 2.41 Mass spectrum of tris-(2-chloro-1-methylethyl) phosphate in a water sample extract obtained from a sewage treatment plant inflow. (Reprint with permission from [121]).

tractions at five different concentrations are given in Table 2.22. The method was also tested for influence of changes in pH (5–9), humic compounds, and detergent concentrations. The recovery rate did not change under these conditions (data not shown). Effects of humic compounds and surfactant concentrations on recovery rate did occur for solid-phase extraction disks in our experiments; thus, in this study the liquid–liquid extraction was used.

Sludge samples were obtained from the loading of trucks that transported the final product from the plant to incineration or final dangerous-waste disposal sites. The respective samples were extracted with ethyl acetate and successively treated with a two-step cleanup procedure utilizing silica sorption and size-exclusion chromatography. The details of this procedure are demonstrated in Sections 2.1.1, 3.2.1, and 3.3.

The standard compound was obtained from Akzo-Nobel as the pure product Fyrol PCF and used without further purification. This product contains at least three isomers with the relative composition 9:3:1, the first being tris-(2-chloro-1-methylethyl) phosphate, while the other constituents probably are isomers with increasing contents of the n-propyl group. The isomeric pattern found in the environment is very similar to the one found in the product Fyrol PCF. Only integrals on the main isomer were used for quantification. The isomeric pattern produced by company Rhodia contains significantly more of the second eluting isomer, while the products from Bayer and BASF exhibit only slight differences in isomeric pattern in comparison to the Akzo product.

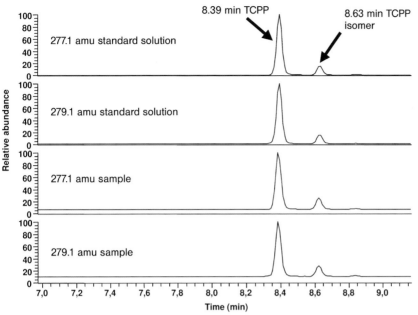

Fig. 2.42 Comparison of retention times of tris-(2-chloro-1-methylethyl) phosphate obtained from a wastewater extract and a standard solution. The second isomer of TCPP that stems from the production process also is displayed, while the third is too small to be visible in this mass fragment chromatogram (measured in SIM mode). (Reprint with permission from [102]).

Table 2.22 Method validation data for the extraction of tris-(2-chloro-1-methylethyl) phosphate from water and sludge, using D_{15} musk xylene as an internal standard.

Medium	Analyzer mass (amu)	Verifier mass (amu)	Recovery rate (%)	SD	RSD	LOQ
Water	277.1	279.1	71	3	4	100 ng L^{-1}
Sludge	277.1	279.1	110	40	35	100 ng g^{-1}

2.4.2.2 Mass Balance Assessment for TCPP in a Sewage Treatment Plant

Tris-(2-chloro-1-methylethyl) phosphate (TCPP) was identified by GC-MS in STP influent and effluent and in the sludge. A spectrum obtained from an inflow water sample is shown in Fig. 2.41. A comparison of retention times in a pure standard solution and a wastewater extract is shown in Fig. 2.42. The last isomer is not visible in the chromatogram, as the peak is very small in environmental samples. Both the spectrum and the retention times matched well with those of an original standard obtained from a Fyrol PCF sample.

The concentrations in the influent of the STP ranged from 200 ng L^{-1} to above 1000 ng L^{-1}. A high day-to-day variation of 59% RSD was detected. As a comparison, the concentrations of some chemical toilet wastes were analyzed. The concentrations in these samples were high, but still the total amount of TCPP introduced by this path to the STP was minute in comparison to the main influent. This day-to-day variation exceeds by far the standard variation of the method itself, which was about 4% RSD. Also, this day-to-day variation was much higher than that obtained for other analytes such as triclosan (average concentration 1200 ng L^{-1}, RSD 7%), which was measured in the same samples ([38], Section 2.4). A situation similar to that of triclosan was experienced for musk fragrances. This clearly indicates that TCPP is introduced into the wastewater stream with a high variation. This result is indicative of discontinuous sources. Domestic washing is supposedly a continuous source in such a large city with a long sewage system, as indicated by the concentration profile for musk fragrances and triclosan that was determined in the same study. Therefore, this cannot be the dominant path of introduction of TCPP to the wastewater. Also, degradation of construction insulation material is a continuous process as long as there are no changes in rainfall, as this may dilute the main inflow or mobilize sediments from the bottom of the sewer system. Thus, these processes, which have been discussed as a pathway of TCPP to the environment, also do not seem to be dominant in this area. The weather was very stable during the sampling of this experiment, with no rainfall at all.

The production of rigid polyurethane foam plates coated with TCPP for flame-retarding reasons is performed by small enterprises, meaning that the product could be released depending on their production cycles. This is supposedly not relevant for this region, as there are no such enterprises identified.

Polyelectrolytes are used to increase the performance of STPs. It was hypothesized that these might also contain flame-retardants. In this STP organic polyelectrolytes are used to increase the speed of sludge production (flocculation). These polymers were analyzed for the presence of flame-retardants, but no organophosphorus flame-retardants were found in the materials used.

On the other hand, the use of treated polyurethane foam on construction sites, especially for the assembly of window frames, etc., may give extremely discontinuous sources. Thus, there was a strong indication that polyurethane foam residues treated with flame-retardants from construction/demolition sites might have been the main source of this compound in this STP. Possible routes of entry for these compounds include unauthorized disposal of materials via the toilets, unauthorized cleaning of tools with water in conventional sinks, cleaning the construction-related dust and dirt in the new buildings, etc. Whether the main source is connected to the processing of rigid polyurethane plates or liquid foam, which makes about 5% of the TCPP market, is unclear at the moment (compare Table 2.23).

A high variability similar to the inflow samples was found in the effluent values (range: 230–610 ng L^{-1}). The concentrations in the effluent were lower than in the influent, indicating some sorption to sludge. The variation of TCPP concen-

Table 2.23 Concentrations of TCPP during a 5-day experiment on an STP. Standard deviations on the day-to-day variability of in-and outflow as well as sludge samples are calculated. From the concentrations in the inflow (in) versus the effluent (eff), elimination (eli) and breakthrough rates (break) were calculated. These calculations were performed as eli = (in − eff)/in (%) and break = 100−eli (%).

Days	Influent (ng L^{-1})	Effluent (ng L^{-1})	Sludge (ng g^{-1})	Elimination (%)	Breakthrough (%)
1	240	240	1700	0	102
2	370	230	2200	38	62
3	1000	610	1300	41	59
4	470	490	1700	0	105
5	480	330	1700	31	69
MW	520	380	1700	21	79
SD	300	170	340		22
RSD (%)	59	43	20		28

trations in the effluent (43% RSD) was slightly smaller than that in the influent. This indicates a smoothing effect of the STP, which is attributed to the fact that in the plant several processes with different time scales (such as activated sludge treatment with an 8-hour hydraulic residence time and anaerobic sludge processing with a 20-day residence time in the anaerobic treatment) overlap.

The concentrations of TCPP in sludge ranged from 1300 ng g^{-1} to 2200 ng g^{-1} (dry weight) during the balancing experiment. These values exhibited less variation than those of the influent (20% RSD). The sewage sludge may thus be better suited for monitoring purposes than wastewater. From the concentrations in the inflow versus the effluent, elimination and breakthrough rates were calculated (Table 2.23). The elimination rates varied from 0% to 41%, leading to breakthrough rates of 51–100% with a day-to-day variation of 28%.

Mass Balance Calculations and Emission Assessment for TCPP in a Sewage Treatment Plant

During the experimental period of five days, about 480 g TCPP came into the plant by means of the influent. About 350 g left the plant by way of the wastewater effluent, while about 480 g were exported with the sludge. These data were obtained with a considerable day-to-day variability, which can be calculated as follows: RSD for influent, 59%; RSD for effluent, 43%; and RSD for sludge, 20%. Apparently, more TCPP was emitted from this STP than was introduced. Either the variability is so huge that real balancing is difficult, or there are transports into the STP that were not included in this dataset. However, in Table 2.28 it is demonstrated that particulate TCPP is an important rout of this compound into waste water treatment plants.

These experiments also lead to the conclusion that TCPP is not degraded in relevant amounts in the STPs under study. This is in accordance with the findings of Kawagoshi et al. [116], who determined TCPP in leachates from solid waste disposal sites. Additionally, they studied whether or not TCPP was degradable in this leachate. The outcome of this study was that this compound was not degradable in incubation experiments lasting for 80 days, though diverse concentrations of oxygen, etc., were employed. Although the manufacturers establish degradation after 20 days experimental time in modified OECD tests, this is also in accordance with these findings, as there is no communal sewage treatment plant in this area that operates with hydraulic retention times longer than 20 hours [117].

To estimate the fraction of TCPP that is released from the products, production data were compared with immission data. Considering that about 7500 t TCPP are used annually by 80 million inhabitants in Germany, this leads to a TCPP consumption of 0.26 g d^{-1} per capita. This corresponds to emissions of 96 g d^{-1} for 340 000 inhabitants, or 0.27×10^{-3} g per capita per day (derived from the Dortmund situation). This means that about 0.1% of the consumed TCPP is emitted to the sewage system.

2.4.2.3 TCPP in Sludge Monitoring

To determine whether the situation in the STP used for the balancing experiment was typical, sewage sludge samples were obtained from 20 different plants in the same German state. These plants were randomly chosen and the concentrations of TCPP in the respective sludge samples were determined. The respective concentrations ranged from 1000 ng g^{-1} to 10 000 ng g^{-1} (dry weight), with one outlier giving 21 000 ng g^{-1}. The respective values are shown in Fig. 2.43. The concentrations in the STP under detailed observation were thus relatively small but quite within the normal range of variations of the other plants. The

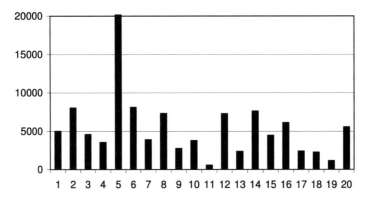

Fig. 2.43 Comparison of TCPP concentrations (ng g^{-1}) in sludges obtained from 20 different STPs in North Rhine-Westphalia. Each data point was generated from a double extraction. The respective plants have been given arbitrary numbers (similar to Fig. 2.9). (Reprinted with permission from [102]).

same effect also was observed for HHCB, AHTN, and triclosan. It should be taken into account, though, that wastewaters from a brewery made up 50% of the wastewater processed in the main plant of this study. Brewery wastewater is not supposed to contain flame-retardants, as they are not used in the brewing process. Thus, the remaining wastewater was diluted 1:1 by the brewery.

High loads of TCPP in sewage sludge are a cause for concern, considering the use of sewage sludge as fertilizers in agriculture. This is especially important because the highest concentration stemmed from a small plant in a less industrialized area of Germany. Sludge from this plant would easily be accepted by farmers. Currently, the sludges of two of the three STPs are deposited at dangerous-waste disposal sites or burnt as dangerous waste, whereas the sludge of one STP is used as fertilizer in agriculture in the state of North Rhine-Westphalia.

2.4.2.4 Evaluation of the TCPP Data

Additionally, the high concentrations in the STP effluent raise some concern about the presence of this compound in surface waters, which may be used as drinking water supplies. An extrapolation to the whole year leads to an estimated annual emission of about 27 kg from this plant. Because it seems that STPS can eliminate only a small fraction of the incoming TCPP if any at all, and it seems there are considerable loads, this may have some implications for STP technology on one hand and the acceptance of TCPP in open usage, e.g., in construction applications, on the other.

2.4.3
Organophosphate Flame-retardants and Plasticizers
in Multi-step Sewage Treatment

This study was performed to obtain information on the behavior of organophosphates in general in wastewater treatment as well as to gain insight into the diverse processes in the respective STPs. Thus, this study was performed on the same plants as mentioned in Section 2.1.1. Wastewater samples from the various steps in two different STPs with preceding and simultaneous denitrification, respectively, were analyzed. Samples were taken before and immediately after the activated sludge tanks as well as from the effluent of the final filtration unit before the treated wastewater was discharged into the receiving river. To obtain information on the efficiency of the STPs in removing organophosphates, it is necessary to study elimination rates over a certain period of time.

Sewage treatment plant A is equipped with a two-stage biological treatment, i.e., two aeration basins and a downstream biological filtration unit (compare Fig. 2.4). Sampling point 1 was located at the main collector prior to the sand trap and the screening plant. The process water from the sludge dewatering was added before the sampling point. The first aeration basin for the raw wastewater, which in this case was highly charged with TOC, was followed by an intermediate settling tank

(IST) before the partially purified water entered the second aeration basin with preceding denitrification. At sampling point 2, samples were collected from the effluent of the IST. Samples of the effluent of the final sedimentation tank (FST) were taken at sampling point 3. The FST was located after the second aeration basin. Before the treated wastewater was discharged to the Rhine, it was filtered through a biological filter. The filter bed consisted of gravel at the bottom and sand at the top. The flow of the treated wastewater and air for the aeration of the filter was from bottom to top. Sampling point 4 was located at the effluent of the STP after the final filtration unit. Sampling points 2 and 3 were chosen to provide information about the elimination of the analyzed compounds at different stages of the wastewater treatment process, whereas sampling points 1 and 4 provided data on the elimination efficiency of the whole process.

2.4.3.1 Materials and Methods

Sewage treatment plant B is a single-stage activated sludge plant with downstream contact filtration. The wastewater flows into the primary settling tank before it enters the aeration basin with simultaneous denitrification. Samples were taken of the influent immediately after the screening plant and the sand trap (sampling point 1) and of the effluent of the PST (sampling point 2). After the biological purification step, the wastewater was separated from the sludge in the final sedimentation tank (FST). Sampling point 3 was located at the effluent of the FST. The wastewater passed through the contact filtration unit before it was fed to the receiving water (the Rhine). The final filter unit was constructed similar to the one in STP A. Sampling point 4 was located at the effluent of the final filtration.

Comparison of the concentrations at sampling points 1 and 2 shows effects of the PST, while the difference between points 2 and 3 demonstrates the efficiency of the aeration basin. Comparison of sampling points 3 and 4 was intended to provide data on effects of the contact filtration.

STP A and STP B are rather large, with wastewater volumes of $109\,000$ m^3 d^{-1} at STP B and of $220\,000$ m^3 d^{-1} at STP A. The corresponding inhabitant equivalent values are $1\,090\,000$ for STP B and $1\,100\,000$ for STP A.

The samples were automatically taken as 24-hour composite samples. The samples were refrigerated at $4\,°C$ during this 24-h interval. They were transported to the laboratory immediately after sampling and extracted within 24 hours after arrival. The samples were generally extracted on the same day by solid-phase extraction using DVB-hydrophobic Speedisks (Mallinckroth Baker, Griesheim, Germany; 45-mm diameter). When it was not possible to extract the organophosphates immediately, the samples were stored at $4\,°C$ overnight.

A solid-phase extraction manifold (IST, Grenzach-Wyhlen, Germany) with PTFE stopcocks and needles was used. Before the extraction, the SPE cartridges were rinsed successively with methyl *tert*-butyl ether (MTBE) and toluene. Afterwards, the disks were conditioned with methanol and water. The water samples were passed through the disks at a flow rate of 200 mL min^{-1} (vacuum). The analytes were successively eluted with MTBE and toluene, and an aliquot of internal standard D_{27} TnBP solution was added to the eluate. The residual water was removed from the organic phase by freezing the samples overnight at –20 °C. The samples were concentrated using a rotary evaporator at 60 °C and 60 mbar to a final volume of 1 mL. Because of matrix interferences, a cleanup of the extracts was necessary, especially for the samples taken of STP influents. For this purpose, a cleanup using silica gel (F60, Merck, Darmstadt, Germany) was established. One gram of dried silica gel (105 °C, 24 h) was put into an 8-mL glass column between two PTFE frits. After conditioning with n-hexane, 1 mL of the sample extract was applied to the column. After a cleaning step with 8 mL n-hexane/MTBE (9:1 v/v), the analytes were eluted twice with 8 mL ethyl acetate. Because not all interferences were eliminated, another internal standard (D_{10} parathion-ethyl) was added at this stage. Afterwards, the volume of the samples was reduced to 1 mL using a rotary evaporator. The solvent was exchanged to toluene and the extract was concentrated to a final volume of 1 mL for GC-MS analysis.

The samples were analyzed on a gas chromatography system with mass spectrometric detection ("Trace" Thermo Finnigan, Dreieich, Germany) equipped with a PTV injector. The PTV (1 µL injection volume) was operated in splitless mode with the following temperature program: 90 °C (0.1 s) → 14.5 °C s^{-1} → 280 °C → 5 °C s^{-1} → 320 °C (5 min) (cleaning phase). The GC separation was performed using a DB-5MS column (J&W Scientific, Folsom, CA, USA) (length: 30 m, i.d.: 0.25 mm, film: 0.25 µm) and the following temperature program: 90 °C (2 min) → 10 °C min^{-1} → 280 °C (15 min), using He (5.0) as carrier gas with a flow of 1.5 mL min^{-1}. The mass spectrometer was used with electron impact ionization with 70 eV ionization energy. The MS was operated in selected ion monitoring (SIM) mode with the detector (photomultiplier) set to a voltage of 500 V.

The different organophosphate esters were detected by means of their mass spectral data and retention times. The method has been validated for quantitative measurements. Recovery rates ranged from 75% to 90% with 5–13% RSD. Full quality data for the method were obtained from three replica extractions of spiked HPLC water at six different concentrations (5, 10, 50, 100, 500, and 1500 ng L^{-1}). The whole set of parameters is given in Table 2.24. Furthermore, municipal wastewater was extracted with the described method and parallel per liquid–liquid extraction with toluene. Both methods gave comparable results.

TCPP and TDCP were obtained from Akzo Nobel (Amersfoort, the Netherlands). These compounds were used without further purification. The technical TCPP gave three peaks at a ratio of 9:3:1. In this study only the main isomer (structure given in Fig. 2.40) was used for determination. TnBP, TiBP, TPP, TCEP, and TBEP were purchased from Sigma-Aldrich (Steinheim, Germany). D_{27} TnBP and D_{10} parathion-ethyl were obtained from Ehrenstorfer (Augsburg, Germany). All solvents were purchased from Merck (Darmstadt, Germany): acetone and ethyl acetate (analytical grade/p.a.), toluene, MTBE, methanol, and n-hexane (residue grade/z.R.). For blank studies, method validation and conditioning of the SPE cartridge water (HPLC grade) from Mallinckrodt Baker (Griesheim, Germany) were used.

2.4.3.2 **Results and Discussion**

STP A Measurements of the influent (sampling point 1) samples showed a considerable day-to-day variation in the concentration of various organophosphorous compounds (range: 570–5800 ng L^{-1} for TCPP and 2400–6100 ng L^{-1} for TBEP). Analysis of the temporal trends revealed that variations on a weekly basis occurred for TCPP only. It seemed that on weekends the load of TCPP in this wastewater treatment plant was lower than on working days. Concentrations in samples from the effluent (sampling point 4) were found to be 1700–6600 ng L^{-1} for TCPP and 290–790 ng L^{-1} for TBEP. These measurements revealed that the non-chlorinated and chlorinated organophosphate esters were eliminated at different rates in wastewater treatment with activated sludge. While elimination of the non-chlorinated organophosphate ester TBEP ranged from 82% to 93%, the chlorinated organophosphates, e.g., TCPP, seemed not to be removed at all. This corresponds with the results of Kawagoshi et al. [116], who found that TCPP could not be degraded in his experiments, and those of Bester [102], who showed that this compound was not eliminated or degraded in a different STP. A comparison of the highest input and output levels of the chlorinated flame-retardants showed that they were of the same order of magnitude. Figures 2.44 and 2.45 give an overview of the measured concentrations of TBEP and TCPP at the different sampling points during the experiment at STP A. The results for all organophosphates are given in Table 2.24.

The concentrations of the different organophosphates at sampling point 2 were of the same order of magnitude as those of the STP influent (point 1). The concentrations were in the range of 600–5900 ng L^{-1} for TCPP and 2300–6100 ng L^{-1} for TBEP in the effluent of the intermediate settling basin (point 2). These data show that the first aeration step did not contribute to the elimination of alkylated organophosphates such as TBEP. At STP A the first biological cleaning step is designed for the fast reduction of dissolved organic carbon (e.g., fats and saccharides) with an average sludge age of one day. Thus, an

Table 2.24 Concentrations of the different organophosphate esters in the influent and effluent of STP A.

Analyte	Max. influent (point 1) (ng L^{-1})	Max. effluent (point 4) (ng L^{-1})	Mean influent (point 1) (ng L^{-1})	Mean effluent (point 4) (ng L^{-1})	Elimination (%)
TiBP	2200	290	1300	160	86 ± 6
TnBP	5500	2300	1200	520	67 ± 16
TCEP	640	410	290	350	None
TCPP	5800	6600	2000	3000	None
TDCP	180	180	100	130	None
TBEP	6100	790	3700	440	88 ± 4
TPP	290	250	130	70	57 ± 24

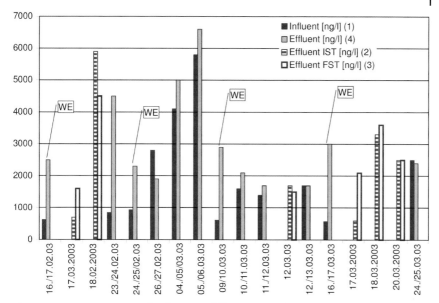

Fig. 2.44 Concentrations (ng L^{-1}) of TCPP at different steps of wastewater purification at STP A (WE = weekend). (Reprint with permission from [124]).

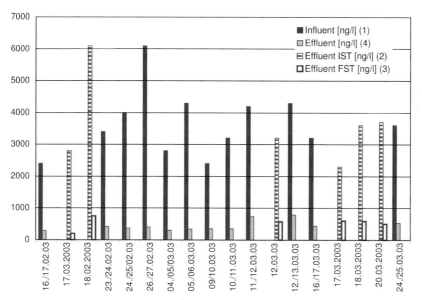

Fig. 2.45 Elimination of TBEP (in ng L^{-1}) at different steps of wastewater treatment in STP A. (Reprint with permission from [124]).

Table 2.25 Concentrations of organophosphate esters at intermediate steps of wastewater treatment at STP A.

Analyte	Max. effluent IST (point 2) (ng L^{-1})	Max. effluent FST (point 3) (ng L^{-1})	Mean effluent IST (point 2) (ng L^{-1})	Mean effluent FST (point 3) (ng L^{-1})	Elimination (%)
TiBP	2300	370	1600	300	79±8
TnBP	4600	670	1100	260	53±25
TCEP	380	430	260	350	None
TCPP	5900	4500	2500	2600	None
TDCP	180	180	100	110	None
TBEP	6100	750	3600	540	84±6
TPP	140	54	93	36	60±20

elimination of xenobiotics by means of biodegradation in this step of the wastewater treatment was not expected. The concentrations of the various organophosphates in the effluent of the final sedimentation (point 3) were of the same order of magnitude as for the effluent (point 4) (1500–4500 ng L^{-1} for TCPP and 250–750 ng L^{-1} for TBEP). The elimination rates calculated on a daily basis for the second aeration basin ranged from 93% to 74% for TBEP, whereas the chlorinated organophosphates (TCPP, TCEP, and TDCP) were not eliminated at all. (The results for the elimination of all organophosphates in STP A are given in Table 2.24). Furthermore, the mean elimination rates for the whole wastewater treatment process were compared to the elimination rates achieved with the second aeration basin. For TiBP, TnBP, and TBEP, the elimination rates for the entire process were slightly higher (2–7%; compare Tables 2.24 and 2.25) than those calculated between the effluent of the intermediate settling (point 2) and the effluent of the final sedimentation (point 3). With regard to the variability of the elimination rates, there was no difference between the rates achieved at sampling points 1 and 4 and those at sampling points 2 and 3. This led to the conclusion that neither the first aeration basin nor the final filtration, but rather the main aeration basin, contributed to the elimination of the non-chlorinated organophosphate esters.

STP B In general, similar data and conclusions were obtained from STP B. A huge day-to-day variability in the concentrations of the different organophosphates was detected. On the other hand, no weekend effect for TCPP was detected in this STP. The concentration ranges were 460–850 ng L^{-1} TCPP and 1800–8000 ng L^{-1} TBEP in the influent and 680–1000 ng L^{-1} TCPP and 65–1200 ng L^{-1} TBEP in the effluent. An overview of all organophosphates is given in Table 2.26. Based on these data the following elimination rates were calculated on a daily basis for the non-chlorinated organophosphates: 72–95% TiBP, 32–76% TnBP, 73–98% TBEP, and 56–87% TPP. As in STP A, the chlorinated organophosphates were not eliminated at all. Figures 2.46 and 2.47 give an over-

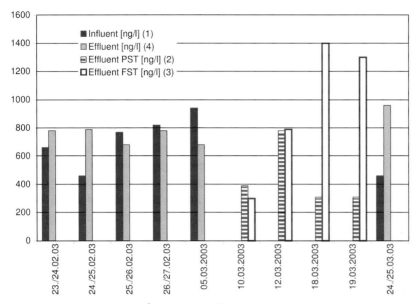

Fig. 2.46 Concentrations (ng L^{-1}) of TCPP at different steps of wastewater purification at STP B. (Reprint with permission from [124]).

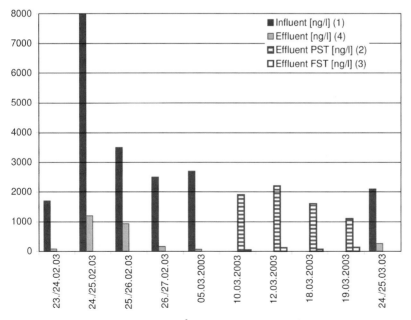

Fig. 2.47 Concentrations of TBEP (ng L^{-1}) at intermediate steps of wastewater treatment at STP B. (Reprint with permission from [124]).

view of the measured concentrations of TBEP and TCPP at the different sampling points during the experiment at STP B. The concentrations of the flame-retardants in the effluent of the primary settling tank (point 2) (260–780 ng L^{-1} TCPP and 1100–1900 ng L^{-1} TBEP) were of the same order of magnitude as in the influent (point 1). This indicated that the primal sedimentation step did not contribute to the elimination of non-halogenated organophosphates (the results for the other organophosphates are displayed in Table 2.27). In the effluent of the final settling basin (point 3), the concentrations were 300–910 ng L^{-1} for TCPP and 46–130 ng L^{-1} for TBEP. From these data, elimination rates for this partial process for each day were calculated as 69–91% for TiBP, 0–40% for TnBP, 88–98% for TBEP, and 9–47% for TPP. Mean values calculated from these elimination rates revealed that the elimination rates for all organophosphates except for TBEP in that part of the cleaning process were slightly lower than for the entire wastewater treatment process. Regarding the variability of the elimination rates, there was no difference between the rates achieved for sampling points 1 and 4 in comparison to the results for sampling points 2

Table 2.26 Concentrations of the different organophosphate esters in the influent and effluent of STP B.

Analyte	Max. influent (point 1) (ng L^{-1})	Max. effluent (point 4) (ng L^{-1})	Mean influent (point 1) (ng L^{-1})	Mean effluent (point 4) (ng L^{-1})	Elimination (%)
TiBP	1500	130	840	78	86±10
TnBP	370	160	260	100	55±15
TCEP	250	470	180	370	None
TCPP	940	1100	650	820	None
TDCP	250	310	110	150	None
TBEP	8000	1200	4000	400	89±9
TPP	140	31	81	20	75±10

Table 2.27 Concentrations of the different organophosphate esters at intermediate steps of wastewater treatment at STP B.

Analyte	Max. effluent PST (point 2) (ng L^{-1})	Max. effluent FST (point 3) (ng L^{-1})	Mean effluent PST (point 2) (ng L^{-1})	Mean effluent FST (point 3) (ng L^{-1})	Elimination (%)
TiBP	1800	270	980	78	83±10
TnBP	400	180	240	100	40±23
TCEP	250	660	310	560	None
TCPP	780	1400	950	820	None
TDCP	120	440	62	310	None
TBEP	2200	130	1700	91	94±4
TPP	100	63	54	38	40±6

and 3. Thus, neither the final filtration step nor the primary sedimentation tank contributed to the elimination of the non-chlorinated organophosphates.

For confirmation purposes, samples from the same STPs from another time period, i.e., October 2002, were analyzed. Similar concentrations and elimination rates were determined.

Particulate Transports in Wastewater

Because considerably more TCPP and other organophosphates were emitted from the STPs with regard to the sum of effluent and sludge data, it was assumed that there was a transport into the STPs that had not been accounted for. As other possibilities were excluded, it was hypothesized that particulate transports were relevant. Because the K_{ow} values of these compounds are relatively small (log K_{ow} 1–4), a particulate transport cannot be due to sorption phenomena on bio-or geogenic material, but it probably would be due to transport of polymeric (plastic) material into the wastewater treatment plant. These polymeric particles may not be fully extracted by the liquid–liquid extraction (LLE) procedure used. Other compounds were included into this study for comparison.

To obtain samples, centrifuged and supernatant water was analyzed separately from the particulate matter. In comparison, uncentrifuged "full" samples were analyzed (Table 2.28). The full and supernatant samples were extracted by the usual LLE with toluene, while the particulate phase was lyophilized and extracted by a vigorous extraction using an accelerated solvent extractor (ASE, ASE 200, Dionex, Idstein, Germany). By this approach, it was considered that those compounds that were sorbed to the (biogeogenic) particles were extracted with the LLE with toluene, while those in which the analytes were dissolved in a polymeric material could be dissolved only by a vigorous accelerated solvent extraction.

In Table 2.28 typical results obtained from a sample from one STP are shown for the particulate and supernatant phase as well as the "total" samples. Addi-

Table 2.28 Particulate and dissolved (supernatant phase) transports of various compounds in the wastewater inflow at STP D (given as ng transport in 1 L) [118].

	T*i*BP	T*n*BP	TCEP	TCPP	TDCP	DEHP	HHCB	AHTN	OTNE
Log K_{ow}		4.0	1.7	2.6	3.8		5.9	5.7	na
Particulate phase	60	0	52	733	0	29386	1652	298	2599
Supernatant phase	68	160	21	92	168	623	2541	312	5629
Sum of the separated phases	128	160	73	825	168	30009	4194	610	8228
"Total" sample	51	165	20	117	199	7476	4971	586	10191
Difference	77	−5	53	707	−31	22533	−777	23	−1963
Relative difference (%)	150	−3	266	603	−16	301	−16	4	−19

tionally, the sums of the supernatant and particulate phases are calculated and compared to the results of the total phase. This is presented as a difference (i.e., sum of the separated phases – total sample). By dividing this result by the "sum of the separated phases" and transforming it into a percentage, the relative difference was obtained. Strong positive results indicate that considerable transports as strongly sorbed particulate phases are relevant.

Considering only the numbers for the particulate and supernatant phases in Table 2.28, it is obvious that particulate transports are relevant for all compounds except TDCP and T*n*BP. However, these transports are less pronounced for HHCB, AHTN, and OTNE, for which the numbers in the particulate phase are smaller than in the supernatant phase. For T*i*BP, TCEP, TCPP, and DEHP, it seems that more of these compounds are transported in the particulate phase than in the dissolved phase.

These findings agree with the comparison of the sum of the separated phases and the total samples. For T*i*BP, TCEP, TCPP, and DEHP, the sum is several hundred percent higher than the total sample, clearly indicating that strong transports of these compounds into the STPs occur with strongly bound fractions of these compounds. These compounds are most probably included in polymeric particles, and thus the current analytical approaches should be altered for total balance assessments.

Another experiment was performed as wet fractionated sieving of an inflow sample. It was expected that small, TOC-rich particles would generally have a higher sorption capacity than the larger particles, and thus it was expected that the small-particle fraction would contain more of the respective xenobiotics. This was true in this case of DEHP, HHCB, AHTN, and OTNE, offering more proof that these compounds indeed underlie normal sorption/desorption equilibria on sediment particles. However, the findings were different for all organo-

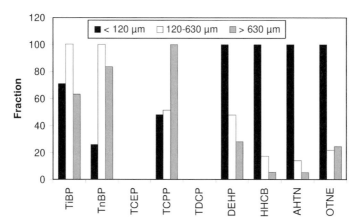

Fig. 2.48 Relative fractions (%) of the respective xenobiotics different size fractions of particulate matter in an STP inflow sample liter. (Data obtained from [118]).

phosphates that were determined in this experiment. These compounds were included even more in the large size particle fractions than in the small size particle fraction (TCPP) or in both particle size fraction (T*n*BP) (Fig. 2.48).

It should thus be considered that compounds such as the organophosphates may enter the STPs as polymeric-bound particulate matter. It is hard to assess these transports, as particulate transports are by no means as homogeneous as dissolved ones. A huge variability is to be expected because the particles most likely do not enter the STPs continuously but are dependend on drift time in the sewer system, hydraulics in the sewer system, as well as complex modes of introduction to the sewer system. Normally, log K_{ow} values are considered to describe the amount of the compound sorbed to particulate matter, but in this case, this is by no means a valid approach.

2.4.3.3 Conclusions

While the chlorinated organophosphates TCPP, TCEP, and TDCP were not eliminated in the wastewater treatment process, the non-chlorinated derivatives T*i*BP, T*n*BP, TBEP, and TPP were partially eliminated in the STPs studied. Additionally, an extremely high day-to-day variability was detected for all organophosphates in these STPs, while compounds such as musk fragrances and triclosan, which were analyzed in the same samples, remained constant throughout the weeks. In the case of TCPP, lower concentrations were detected on weekends in one STP. Elimination occurred mostly in the main aeration basin, which may be attributed to sorption to sludge as well as to biodegradation processes.

At both STPs the efficiency of the cleaning process for the organophosphate esters was comparable. Thus, the type of construction of the STP was not relevant for the elimination of these substances.

2.4.4
Organophosphorus Flame-retardants and Plasticizers in Surface Waters

As was demonstrated in Sections 2.4.2 and 2.4.3, organophosphates, especially the chlorinated compounds, are not eliminated well in sewage treatment. It was thus decided to study their dispersion in surface water in comparison to other organophosphates.

The experimental area of the Ruhr was chosen because this river supplies several million inhabitants of the Ruhr megalopolis with water for drinking-water extraction. Rivers such as the Lippe, the Rhine, and the Elbe were used for confirmation.

2.4.4.1 Materials and Methods
The Ruhr is a small river with 2200 million m^3 water flow annually near the mouth (Hattingen). The spring is located in the moderately populated Sauerland area. It passes several lakes until it reaches the industrial Ruhr area, in

which it feeds into the purification plants that supply drinking water to about 5 million inhabitants. After the river has passed the plants located near the cities of Dortmund (near sample station 50), Bochum (in the vicinity of sample station 56), and Essen (near sample station 63), it reaches the Rhine (compare Fig. 2.11 and Table 2.7). Several other tributaries (such as the Möhne River) are also used to control the water flow in the Ruhr in a way that the water extraction plants can operate continuously.

During the sampling in September 2002, the flow of the river was about 25–28 $m^3 s^{-1}$, which was significantly lower than average. This was due to the fact that the sampling period took place in the middle of a dry period with no rainfall at all. This period was chosen to enable us to perform back calculations with software packages such as EUSIS or GREAT-ER. Rainfall can hardly be calculated, as the documentation on rainfall is done with too little spatial resolution. The exact location and a characterization of sampling sites are shown in Table 2.7.

The water samples were immediately extracted with 10 mL toluene after adding an aliquot of internal standard solution (D_{27} TnBP). The extraction (30 min) was performed by vigorous stirring with a Teflonized magnetic stirrer. After a sedimentation phase of 20 min, the organic phase was separated from the aqueous one and the residual water was removed from the organic phase by freezing the samples overnight at $-20\,°C$. The samples were concentrated with a rotary evaporator at $60\,°C$ and 60 mbar to 1 mL. They were analyzed by GC-MS (Thermo-Finnigan Trace) equipped with a PTV injector. The PTV (1 µL injection volume) was operated in PTV splitless mode with the following temperature program: $90\,°C$ (0.1 s) → $14.5\,°C\ s^{-1}$ $280\,°C$ (1.0 min) → $5\,°C\ s^{-1}$ → $320\,°C$ (5 min) (cleaning phase). The GC separation was performed utilizing a DB-5MS column (J&W Scientific) (length: 30 m; i.d.: 0.25 mm; film: 0.25 µm) and a temperature program of $90\,°C$ (2 min) → $10\,°C\ min^{-1}$ → $280\,°C$ ($15\,°C\ min^{-1}$) utilizing He with

Table 2.29 Quality assurance data for the determination of organophosphates from water. Retention time, analytical ions in atomic mass units (amu), recovery rate (rr), limit of quantification (LOQ) (ng L^{-1}), working range of the respective compounds, using D_{27} TnBP as internal standard.

Compound	Retention time (min)	Analytical ion (amu)	Verification ion (amu)	Recovery rate (%)	RSD (%)	LOQ (ng L^{-1})
TiBP	10.56	155	211	107	12	6.3
TnBP	12.17	155	211	98	19	10
TCEP (LLE)	13.50	249	251	31	33	20
TCEP (SPE)	dto	dto	dto	67	15	12
TCPP	13.85	277	279	101	14	4.9
TDCP	18.92	381	379	95	3	14
TPP	19.61	325	326	93	27	10
TBEP	19.57	199	125	89	19	6.4

1.5 mL min^{-1} as carrier gas. The mass spectrometer was operated at 500 V on the photomultiplier and about 78 ms dwell time in SIM mode. The whole set of parameters is given in Table 2.29.

Because TCEP was not recovered very well by liquid–liquid extraction (LLE), an alternative solid-phase extraction (SPE) utilizing DVB-hydrophobic Speedisks (Mallinckroth Baker, Griesheim, Germany) with a diameter of 45 mm was established.

TCPP and TDCP were a gift from Akzo-Nobel (Amersfoort, the Netherlands). The compounds were used as received without further purification. Technical TCPP gives three peaks at a ratio of ∼9:3:1. The quantification was performed by referring to the first peak only. T*n*BP, T*i*BP, TPP, TCEP, and TBEP as well as humic acid sodium salts were purchased from Sigma-Aldrich (Steinheim, Germany). D$_{27}$ T*n*BP was obtained from Ehrenstorfer (Augsburg, Germany). All solvents were purchased from Merck (Darmstadt, Germany): acetone (analytical grade/p.a.), toluene (residue grade/z.R.), and Triton X-100 (p.a. grade). For blank and recovery studies, water from a Millipore apparatus (Milli-Q Gradient with an Elix 3 purification unit; Millipore, Schwalbach, Germany) was used.

2.4.4.2 Results and Discussion

Organophosphates such as TCPP were identified by mass spectra and retention times in surface waters. In Fig. 2.41 the mass spectrum of tris-(2-chloro-1-methylethyl) phosphate obtained from a water sample is shown. The chromatograms obtained for surface water look very similar to those obtained for wastewater (Fig. 2.43). For quantitative measurements the extraction procedure was validated. It gave recovery rates of 89% to 107% with 11–29% RSD (see Table 2.29). Only TCEP was not recovered well by this LLE procedure; in addition, standard deviations were high, and thus all presented data for TCEP are considered to be indicative data rather than "true" data. The SPE method gave good recoveries for TCEP, however. At some places this method was employed in parallel to the LLE procedure, which gave similar results. Full quality data of the method obtained from three replicate extractions at eight different concentrations (2, 10, 20, 100, 200, 1000, 2000, and 10 000 ng L^{-1}) are given in Table 2.29. Both methods were also tested for influence of changes in pH (5–9), humic compounds with concentrations varying from 5 mg L^{-1} to 5000 mg L^{-1}, and detergent (Triton X-100) concentrations varying from 50 µg L^{-1} to 5000 µg L^{-1}. The recovery rates for the LLE procedure did not change under these conditions, while the highest concentrations of detergents gave lower recovery rates for all analytes on the SPE extractions.

Organophosphorus Flame-retardants

In Fig. 2.49 the distribution of TCPP concentrations in the Ruhr, in its main tributaries, and in several STP effluents is shown. The concentrations of TCPP in the Ruhr varied between 20 ng L^{-1} and 200 ng L^{-1}. All STPs that were sampled contributed considerably to the load of TCPP in the river, as typical concentrations of 50–400 ng L^{-1} were analyzed in the effluents. It is no surprise

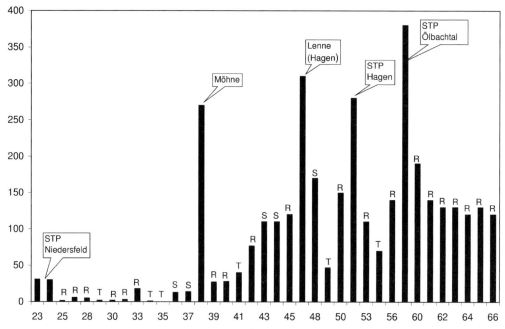

Fig. 2.49 Distribution of TCPP (ng L^{-1}) in Ruhr River water, in the Möhne and Lenne tributaries, and in effluents from STP Niedersfeld, STP Hagen, and STP Bochum-Ölbachtal. Samples 57–59 are displayed as average. R: riverine samples; T: tributary samples. (Reprinted with permission from [121]).

that samples from upstream of STP Niedersfeld (sample stations 25 and 26) were very low in concentration, as no inflow whatsoever is known to occur between there and the spring of the river, which is a small creek at that place. It is slightly surprising that high concentrations were measured in the tributary Möhne (300 ng L^{-1}), which is generally supposed to be little affected by STP effluents. On the other hand, the tributary Lenne often has contamination patterns that are connected to STP effluents. The STPs Hagen and Bochum-Ölbachtal introduced high concentration of TCPP into the Ruhr. The tributary Volme, on the other hand (station 55), showed rather low ones. Interestingly enough, a high concentration of 100 ng L^{-1} was reached at station 42 (upstream of Fröndenberg), before the Ruhr entered the densely populated and industrialized Ruhr area. This concentration stayed around 100–150 ng L^{-1} until the river passed Essen and Mülheim and a few kilometers before it reached its mouth at the Rhine in Duisburg. Via this route it passes several lakes (e.g., Lake Kemmnaden, stations 61–63). The lakes do not seem to change the concentrations, although they are generally supposed to have a "cleansing effect."

It is interesting to note that the main STPs sampled gave quite different emissions of TCPP per capita (Niedersfeld: 11 µg d^{-1}; Menden: 34 µg d^{-1}; Hagen-Fley: 31 µg d^{-1}; and Bochum-Ölbachtal: 220 µg d^{-1}).

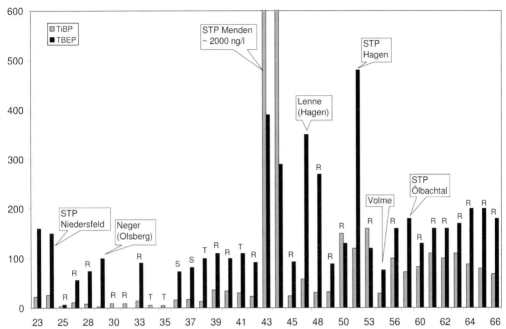

Fig. 2.50 Distribution of TBEP and TiBP (ng L^{-1}) in Ruhr water, in the Möhne and Lenne tributaries, and in effluents from STP Niedersfeld, STP Menden, and STP Hagen. Samples 57–59 are displayed as average. RR: riverine samples; T: tributary samples. (Reprint with permission from [121]).

Data on TCEP (indicative) and TDCP basically showed a similar distribution, again with high values in the tributary Möhne, especially for TDCP. However, all concentrations of these two compounds were lower than those of TCPP. Final concentrations of TCEP and TDCP were about 50 ng L^{-1}. The concentrations in STP effluents were 5–130 ng L^{-1} for TCEP and 20–120 ng L^{-1} for TDCP. The concentrations of especially TCPP are much too high to be caused by electronic equipment alone, as in the experiment of Carlsson [97]. Thus, applications that involve high amounts of TCPP, such as polyurethane foam plates or liquid spray foam, are probably more relevant.

Organophosphorus Plasticizers
In the same experiment, organophosphate plasticizers such as TiBP, TnBP, TBEP, and TPP were analyzed because they appeared in the same chromatograms and gave good recoveries as well (Table 2.29). The concentrations of these compounds were similar to TCPP (10–200 ng L^{-1}) but in some cases were higher (compare Fig. 2.50). The concentrations of TBEP rose to nearly 500 ng L^{-1} in several STP effluents. The concentrations of TiBP reached as high as 2000 ng L^{-1} in the effluent of STP Menden near Fröndenberg. For these compounds only the direct STP discharges and the tributary Lenne were relevant. Other tributaries, such as the river

Möhne, which was a major source for the chlorinated compounds, were not dominant for the plasticizers. Interestingly enough, high concentrations of TBEP appeared at station 36 (effluent of STP Meschede) and were then stable in the river itself, staying at 100 ng L^{-1} for some time. Again, tributary Volme (station 55) was relatively uncontaminated. During the passage of the densely populated Ruhr area, the concentrations of TBEP rose until they reached a stable level of about 200 ng L^{-1} near its mouth. STP Bochum-Ölbachtal did not contribute to the contamination of the river with TBEP.

TiBP, on the other hand, exhibited low concentrations ($<$ LOQ to 25 ng L^{-1}) for all samples upstream of the river, including for STP effluents. The concentrations increased near Arnsberg (station 37), and huge concentrations were introduced from STP Menden (stations 43 and 44), thus leading to elevated concentrations (150 ng L^{-1}) in the Ruhr near Schwerte (station 50), where raw water for purification for the drinking water supply of the city of Dortmund is abstracted from the river. Neither the tributary Lenne nor the tributary Volme showed higher concentrations of TiBP than the Ruhr River itself. Again, STP Ölbachtal did not exhibit elevated levels in the effluent. The concentrations of TiBP were stable at around 100 ng L^{-1} until the river reached Mülheim.

TnBP and TPP showed lower concentrations in the whole experiment. TnBP reached its highest concentration of 110 ng L^{-1} upstream between Olsberg and Meschede (stations 33–35 and 37); otherwise, the concentrations were 30–40 ng L^{-1}, with the highest concentrations downstream near the mouth.

The highest concentration (40 ng L^{-1}) of TPP was found in a harbor for leisure boats in Lake Kemnaden (station 60). Some STP effluents had concentrations of 10–30 ng L^{-1}. Thus, the concentrations of TPP generally were low in comparison to the other organophosphates.

A temporal comparison was performed by comparing samples from September to samples from July. These data are shown in Table 2.30. Of course, the

Table 2.30 Comparison of the concentration (ng L^{-1}) of organophosphates during two sampling campaigns in the Ruhr River in July and September 2002.

Compound	TiBP	TnBP	TCEP	TCPP	TDCP	TBEP	TPP
Sample location							
46 (July)	70	70	180	100	$<$LOQ	870	60
(September)	58	13	45	310	27	350	17
50 (July)	50	60	190	130	$<$LOQ	$<$LOQ	$<$LOQ
(September)	150	26	81	150	41	130	12
57 (July)	70	70	300	280	$<$LOQ	290	60
(September)	100	24	48	140	46	160	13
60 (July)	80	60	250	290	$<$LOQ	230	80
(September)	83	26	58	190	57	130	39
61 (July)	160	130	$<$20	230	$<$LOQ	$<$100	$<$10
(September)	110	34	45	140	42	160	10

hydrodynamic situation was not exactly the same, and thus the results showed some differences. The concentrations of TiBP and TCPP are very similar in both sets; higher variance was obtained for TnBP, and TCEP was analyzed with a high standard deviation. This may be the main reason why these values showed some variance. TDCP, TBEP, and TPP were near the detection limits, especially in the first sampling series.

As a comparison with other rivers, five samples from the Rhine and a duplicate sample from the Lippe were analyzed. Both of these rivers are supposed to be less protected than the Ruhr. The results for flame-retardants were TCPP 80–100 ng L^{-1} (Rhine) and 100 ng L^{-1} (Lippe); TDCP 13–36 ng L^{-1} (Rhine) and 17n g L^{-1} (Lippe). The following concentrations were measured for the plasticizers: TiBP 30–50 ng L^{-1} (Rhine) and 100 ng L^{-1} (Lippe); TnBP 30–120 ng L^{-1} (Rhine) and 30 ng L^{-1} (Lippe); TBEP 80–140 ng L^{-1} (Rhine) and 130 ng L^{-1} (Lippe). It seems the high standard of protection that is often claimed for the Ruhr was not very effective concerning organophosphates. Tentative samples from the Mulde (an Elbe tributary) exhibited similar concentrations, though the pattern was different because in those samples TCEP was detected in higher concentrations than TCPP. These concentrations were in the same range as stated by Aston [100] for Japanese (17–350 ng L^{-1}), Canadian (about 10 ng L^{-1}), and U.S. rivers (570 ng L^{-1}) (all TCEP data). In Spain 10–900 ng L^{-1} TiBP and about 350 ng L^{-1} TCEP was detected by Barcelo et al. [119]. Prösch et al. [115] detected TCEP and TCPP concentrations in STP effluents varying from 14 ng L^{-1} to 1660 ng L^{-1} and from 18 ng L^{-1} to 26 000 ng L^{-1}, respectively. Prösch et al. discussed a connection to textile production and textile washing as well as industrial point sources in the sewer system. This hypothesis may explain the extraordinarily high concentrations determined in these samples.

From the data in the study presented in this section, it seems that STPs do in some cases emit specific patterns of organophosphates. With the approach demonstrated in more detail in Section 2.2.3 for triclosan, half-lives of the respective organophosphates were assessed for in-river elimination. In Table 2.31 the respective data are displayed. The elimination rates for most compounds are insignificant considering the residence time of about 6 days in this part of the river. The methodology of this approach is laid out in more detail in Section 2.2.3. The precision of the assessment of the half-lives is determined mostly by the

Table 2.31 Assessment of in-river elimination rates of organophosphates in the Ruhr.

Compound	Elimination rate constant (d^{-1})	Half-life (d)
TnBP	0.025	>27
TiBP	0.011	>63
TCPP	0.012	>58
TDCP	0.023	>31
TPP	0.051	>14

precision of the analytical data; thus, the given half-lives are minimum values and could also be much higher.

Mass Flow Considerations for Organophosphates in Surface Waters

Because the Ruhr is among Europe's most important rivers for drinking water, which is kept as clean as possible with low sewage discharges in comparison to other rivers, it was surprising to find these compounds at all. Among the flame-retardants TCPP is the most prominent, which corresponds well with the current sales figures, as industries are phasing out TCEP and TDCP. The industry states that in 1998 about 7500 t TCPP, 750 t TDCP, and 100 t TCEP were sold [108]. The sales are supposed to have shifted further to TCPP in the meantime. Thus, it must be concluded that the origin of TCPP is not the decomposition of old buildings but rather current activities. Otherwise, the concentrations of TCEP and TDCP, which dominated the European market until 1999, would have been higher in relation to TCPP. TCPP most likely stems from current construction activities, either by the handling of rigid foam plates or by usage of liquid spray foam. The concentration pattern found in the Ruhr is in some part a universal background, as TCPP reaches the Ruhr from a multitude of sources. On the other hand, some sources are exceptionally high, leading to the assumption that peak or point sources (possibly large-scale construction sites) are relevant as well. An estimate of transport for organophosphates leads to the assumption that about 300 kg TCPP and about 100 kg each of TDCP and TCEP are transported from the Ruhr to the Rhine annually, which would correspond to 0.005% of the annual consumption in Germany or $\sim 0.1\%$ of the product presumably consumed in the Ruhr region.

The Rhine holds concentrations of 10–90 ng L^{-1} of organophosphates, leading to annual transports of these compounds of 0.7 t (TCEP) to 7 t (TCPP and TBEP) with the Rhine via the Netherlands to the North Sea. An overview is given in Table 2.32.

The situation of the plasticizers is somewhat similar. TBEP and TiBP are the most relevant compounds in the Ruhr system. Though these compounds are omnipresent, there are some relevant point sources. In this case the point

Table 2.32 Transports of organophosphates in the Rhine to the North Sea via the Netherlands.

Compound	Concentration in the Rhine (ng L^{-1})	Annual transport (t/a)
TCPP	90	6.7
TiBP	40	3.0
TnBP	40	3.0
TCEP	10	0.74
TDCP	14	1.0
TEBP	90	6.7
TPP	10	0.7

sources are diverse, and the same emission patterns are not found for the other compounds that were analyzed. An estimate of transport leads to the assumption that about 300 kg TBEP and 200 kg TiBP are transported into the Rhine annually.

Generally, it should be considered that similar concentrations would be found in surface waters all over Europe, as we found these compounds in several rivers of different regions in Germany. Similar concentrations (20–200 ng L^{-1} TCEP and 200–700 ng L^{-1} TBEP) have been published by Fries and Püttmann [114].

At the moment this does not necessarily mean harm to the population of the Ruhr area, as most of these compounds are probably successively eliminated by the water purification plants. However, some reasons for concern regarding human health remain. In Canada some of these compounds (0.6–12 ng L^{-1} TBP, 0.3–9.2 ng L^{-1} TCEP, 0.2–1.2 ng L^{-1} TDCP, 0.9–75 ng L^{-1} TBEP, 0.3–2.6 ng L^{-1} TPP) have been found in drinking water [120]. Little is documented on the purification processes used on these samples, though. On the other hand, the consumer has to pay for the installation and maintenance of the considerable efforts that the water suppliers make to eliminate xenobiotics from the raw water from a river like the Ruhr or the Rhine. Additionally, TCPP is assumed to be a carcinogen; thus, it is a major issue for the wildlife of the river as well.

The applications that are dominant at the moment should be examined for potential emissions of the respective compounds. It is possible that simple changes in the installation or application of rigid polyurethane foam plates and liquid spray foam can reduce the concentrations in relevant rivers considerably. This could reduce costs for consumers and may improve the evaluation of major rivers with regard to the Water Framework Directive of the EU [4]. Because Kawagoshi [116] found that TCPP could not be degraded in his experiments or in STPs [102, 124], improving the degradation or elimination power of STPs will probably be hard to achieve, as well as quite costly, for reducing the concentrations of TCPP and other organophosphates in surface waters.

2.4.5
Organophosphates in Drinking Water Purification

Because several organophosphate flame-retardants and plasticizers were demonstrated not to be eliminated in considerable amounts from wastewater treatment plants and thus were present in large amounts in surface waters, one reason for concern is the production of drinking water from surface water or enriched groundwater as is presently the case in the Ruhr megalopolis. In this region about 7 million inhabitants are served with water stemming either directly or indirectly from the Ruhr.

For this study three waterworks that purify surface water from the Ruhr were chosen. This river has been protected since the third decade of the nineteenth century by STPs and, additionally, it is preferable to introduce wastewater not into the Ruhr itself but into other rivers such as the Emscher whenever possible. However, previous studies have shown that it is still affected by STP effluents, as the treated wastewater of about two million inhabitants is discharged into this river [121]. During the summer months, the Ruhr contains up to 30% wastewater.

Le Bel et al. [120] detected organophosphate esters (0.3–9.2 ng L^{-1} TCEP and 0.5–11.8 ng L^{-1} TnBP) in drinking water samples from six eastern Ontario water treatment plants. Elimination during drinking water purification was not observed, though.

A study on the persistence of pharmaceutical compounds and other organic wastewater contaminants in a conventional drinking water treatment plant in the U.S. showed that the applied treatment processes were not effective in removing TBP, TBEP, TCEP, and TDCP [122]. Heberer et al. [123] described the production of drinking water from highly contaminated surface waters by applying mobile membrane filtration units. In this study elimination rates observed for TCEP and TCPP were >97.2% and >98.9%, respectively. The objective of the current work was to study the efficiency of different treatment steps in removing organophosphate esters from surface water for drinking water purification. Therefore, the elimination of these substances was studied in three different waterworks with different treatment processes in the Ruhr area.

2.4.5.1 Materials and Methods

Selected Waterworks and Sample Locations
In the Ruhr megalopolis the combination of different treatment processes depends on the quality of the raw water used for drinking water purification. This means that purification plants that are located downstream of the highly populated and industrialized area of the Ruhr megalopolis have to use additional treatment processes to obtain drinking-water quality. In this study the elimination efficiency of three waterworks was compared. Waterworks A (see Fig. 2.51) is located in a more or less rural area upstream of the highly industrialized area. This water treatment facility (A) is subdivided into two waterworks. After the water purification, the treated water of both waterworks is fed to the public water supply. One of the two waterworks is equipped with gravel pre-filters and main filters (a combination of biologically active slow sand filters and underground passage), whereas the other one uses bank filtration and slow-process sand filtration combined with underground passage.

Waterworks B is located near the mouth of the Ruhr. In this water treatment facility, biologically active slow-process sand filtration with underground passage is combined with secondary treatment processes such as ozonization, multilayer and activated carbon filtration, and UV irradiation for disinfection purposes (for details, see Fig. 2.51).

For drinking water purification in waterworks C, the same treatment processes as in waterworks B are used. However, they occur in a different order.

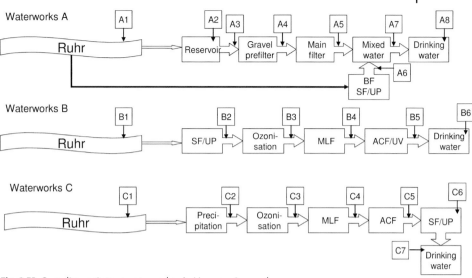

Fig. 2.51 Sampling points at waterworks. A: Hengsen-Lappenhausen; B: Mülheim-Styrum Ost; C: Mülheim-Styrum West. A, B, C: sampling points; SF: sand filtration; UP: underground passage; BP: bank filtration; MLF: multilayer filter; ACF: activated carbon filter; UV: UV irradiation; Pre/Floc: precipitation and flocculation. (Reprint with permission from [124]).

Additionally, the raw water is treated with aluminum salts for precipitation and flocculation (see Fig. 2.51).

Except for samples of the Ruhr and from the reservoir at waterworks A, all samples were taken at sampling points used for routine monitoring at the respective water treatment facilities. In each case the sample volume was more than 2 L that were divided into two 1-L samples for two replica extractions. The Ruhr and the reservoir of waterworks A were sampled near the inflow of the waterworks and the pre-filter, respectively. The samples were taken at the same time but not according to the supposed residence time. Thus, waterworks A was sampled over a period of five days to study the continuity of the elimination efficiency, as the contact time for the slow-process sand filtration and underground passage was 12–15 days. From each sampling point, one grab sample was collected per day.

Chemicals

TCPP and TDCP were obtained from Akzo Nobel (Amersfoort, the Netherlands). These compounds were used without further purification. The technical TCPP gives three peaks at a ratio of $9:3:1$. In this study only the main isomer was used for determination. TnBP, TiBP, TPP, TCEP, and TBEP were purchased from Sigma-Aldrich (Steinheim, Germany). D_{27} TnBP was obtained from Ehrenstorfer (Augsburg, Germany). D_{15} TPP was synthesized from D_6 phenol and phosphoroxytrichloride. All solvents were purchased from Merck (Darmstadt, Germany): acetone (analytical grade/p.a.), ethyl acetate (analytical grade/

p.a. and residue grade/Suprasolv), and toluene and methanol (residue grade/Suprasolv). For blank studies and method validation, water (HPLC grade) from Mallinckroth Baker (Griesheim, Germany) was used.

Analytical Methods

All samples were collected in glass bottles and stored at 4 °C when it was not possible to extract them immediately. The storage time was no longer than 48 h. The results were obtained from two replica extractions of each sample by means of liquid–liquid extraction (LLE). One liter of the samples was extracted with 10 mL toluene after adding an aliquot (100 μL) of internal standard solution containing D_{27} TnBP (1.8 ng μL^{-1}) and D_{15} TPP (1.01 ng μL^{-1}). The extraction (30 min) was performed by vigorous stirring with a Teflonized magnetic stirrer. After a sedimentation phase of 20 min, the organic phase was separated from the aqueous one and the residual water was removed from the organic phase by freezing the samples overnight at –20 °C. The samples were concentrated with a concentration unit (Büchi Syncore, Büchi, Essen, Germany) at 60 °C and 60 mbar to 1 mL. For blank studies, water (HPLC grade, Baker Griesheim, Germany) was treated under the same conditions as water samples. None of the selected organophosphates was detected in blank samples except for TPP. The blank value was traced back to one batch ethyl acetate p.a. that was used for the cleaning of the glass bottles. Afterwards, ethyl acetate (residue grade) was applied for cleaning purposes. For each set of samples, instrumental and procedural blanks were analyzed.

The samples were analyzed on a gas chromatography system with mass spectrometric detection (DSQ Thermo Finnigan, Dreieich, Germany) equipped with a PTV injector. The PTV was operated in large-volume injection (LVI) mode (40 μL injection volume) with a sintered glass liner (SGE) with the following temperature program: 115 °C (0.4 min, 130 mL min^{-1} He 5.0) → 12 °C s^{-1} (splitless) → 280 °C (1.2 min) → 1 °C min^{-1} → 300 °C (7 min) (cleaning phase).

The GC separation was performed using a DB5-MS column (J&W Scientific, Folsom, CA, USA) (length: 15 m; i.d.: 0.25 mm; film: 0.25 μm) and the following temperature program: 100 °C (1 min) → 30 °C min^{-1} → 130 °C → 8 °C min^{-1} → 220 °C → 30 °C min^{-1} → 280 °C (7 min). The mass spectrometer was used with electron impact ionization with 70 eV ionization energy. The MS was operated in selected ion monitoring (SIM) mode. Mass fragments that were used for quantification are given in Table 2.33.

The different organophosphate esters were detected by means of their mass spectral data and retention times. For quantitative measurements the method was validated. Recovery rates ranged from 28% to 128%, with 7–19% RSD for the LLE. Full quality data for the method were obtained from three replica extractions of spiked HPLC water at nine different concentrations in the range of 1 ng L^{-1} to 10 000 ng L^{-1} for the LLE. The whole set of parameters is given in Table 2.33. Because TCEP was not recovered well by LLE, a solid-phase extraction (SPE) method was developed for the determination of this substance from surface water. A comparison of both methods gave the same results as samples taken from the Ruhr in 2002 (for details, see Ref. [121]).

Table 2.33 Quality assurance data for LLE of organophosphates from drinking water (40 μL injection volume) and chemical description of mass fragments used for quantification. IS: internal standard. This method was used to determine organophosphates in drinking water processing and for multicomponent analysis for samples from the North Sea and Lake Ontario.

Analyte	Analytical ion (amu)	Verifier ion (amu)	Recovery rate (%)	RSD (%)	LOQ (ng L–1)	IS
TiBP	211 $[M\text{-}C_4H_9]^+$	155 $[M\text{-}C_8H_{18}]^+$	128	13	3	$D_{27}TnBP$
TnBP	211 $[M\text{-}C_4H_9]^+$	155 $[M\text{-}C_8H_{18}]^+$	100	11	1	$D_{27}TnBP$
TCEP	249 $[M\text{-}^{35}Cl]^+$	251 $[M\text{-}^{35}Cl]^+$	28	12	0.3	$D_{27}TnBP$
TCPP	277 $[M\text{-}CH_2{}^{35}Cl]^+$ $[C_8H_{16}{}^{35}Cl_2O_4P]^+$	279 $[M\text{-}CH_2{}^{35}Cl]^+$ $[C_8H_{16}{}^{35}Cl^{37}ClO_4P]^+$	92	10	1	$D_{27}TnBP$
TDCP	379 $[M\text{-}CH_2{}^{35}Cl]^+$	381 $[M\text{-}CH_2{}^{35}Cl]^+$	108	13	1	D_{15} TPP
TBEP	199 $[M\text{-}C_{12}H_{26}O_2]^+$	299 $[M\text{-}C_6H_{13}O]^+$	103	7	3	D_{15} TPP
EHDPP	251 $[M\text{-}C_8H_{17}]^+$	362 $[M]^{+\bullet}$	94	11	0.1	D_{15} TPP
TPP	325 $[M\text{-}H]^+$	326 $[M]^{+\bullet}$	101	14	0.3	D_{15} TPP
HHCB	243 $[M\text{-}Me]^+$	258 $[M]^{+\bullet}$	87	13	0.1	D_{15} musk xylene
HHCB-lactone	257 $[M\text{-}Me]^+$	272 $[M]^{+\bullet}$	100	9	0.03	D_{15} musk xylene
Musk ketone	279 $[M\text{-}Me]^+$	294 $[M]^{+\bullet}$	90	23	0.01	D_{15} musk xylene
Musk xylene	282 $[M\text{-}Me]^+$	297 $[M]^{+\bullet}$	95	9	0.01	D_{15} musk xylene
AHTN	243 $[M\text{-}Me]^+$	258 $[M]^{+\bullet}$	105	19	0.1	D_{15} musk xylene
Triclosan	288 $[^{35}Cl_3\text{-}M]^{+\bullet}$	290 $[^{35}Cl_2{}^{37}Cl_1\text{-}M]^{+\bullet}$	88	9	0.1	D_{15} musk xylene
Methyl triclosan	302 $[^{35}Cl_3\text{-}M]^{+\bullet}$	304 $[^{35}Cl_2{}^{37}Cl_1\text{-}M]^+$	102	12	0.1	D_{15} musk xylene

2.4.5.2 Results

Chlorinated Organophosphates

Table 2.34 gives an overview of the concentrations of TCEP, TCPP, and TDCP in waterworks A at the respective sampling points. Because samples were taken over a period of five days, the concentrations are additionally given as mean values. The amounts of TCPP were reduced from 54 ng L^{-1} in the Ruhr to 2.9 ng L^{-1} in the finished water (95% elimination), those of TDCP from

Table 2.34a Concentrations of the selected chlorinated organophosphates at different treatment steps at waterworks A (PF: pre-filter; MF: main filter; UP: underground passage; MW: mixed water; FW: finished water).

Analyte	Ruhr [ng/L]	Reservoir [ng/L]	PF inflow [ng/L]	MF inflow [ng/L]	MF effluent [ng/L]	UP effluent [ng/L]	MW [ng/L]	FW [ng/L]	Day
	A1	A2	A3	A4	A5	A6	A7	A8	
TCPP	47	49	45	41	<1	<1	<1	<1	1
	55	50	57	46	<1	<1	2.5	1.2	2
	47	52	45	44	12	<1	2.9	3.4	3
	57	59	52	48	3.5	<1	6.9	4.1	4
	65	59	57	51		<1	10		5
Mean	54	54	51	46	7.8	<1	5.6	2.9	
TDCP	10	12	11	11	3.2	1.4	1.3	1.2	1
	15	12	13	15	1.3	1.4	1.3	1.9	2
	8.6	10	10	11	6.5	1.3	2.9	2.4	3
	11	12	10	9.8	2.4	1.1	3.4	2.4	4
	18	14	14	14		1.1	4.0		5
Mean	13	12	12	12	3.4	1.3	2.6	2.0	
TCEP	12	14	13	12	0.61	0.65	1.3	1.2	1
	130	26	23	47	0.56	0.64	1.3	1.9	2
	13	15	13	14	2.8	0.70	2.9	2.4	3
	20	51	21	18	1.4	0.56	3.4	2.4	4
	32	22	23	23		0.51	4.0		5
Mean	41	26	19	23	1.3	0.61	2.6	2.0	

13 ng L^{-1} to 2.0 ng L^{-1} (85% elimination), and those of TCEP from 41 ng L^{-1} to 2.0 ng L^{-1} (95% elimination) in the complete treatment process. Because the respective concentrations of the chlorinated organophosphates in the influent of the pre-filter and the influent of the main filter were constant in this experiment, the pre-filter did not contribute to the elimination of these substances. Moreover, Table 2.34 shows that the concentrations of TCEP in the Ruhr, the reservoir, and the influents of the pre-filter and the main filter exhibit significant variability, whereas they were almost stable for TCPP and TDCP. The concentrations of TCEP in the Ruhr ranged from 13 ng L^{-1} up to 130 ng L^{-1}. This variance is also reflected in the values measured in the reservoir and in the inflows of the pre-filter and the main filter.

Table 2.34a demonstrates that the concentrations of chlorinated organophosphates showed a significant day-to-day variance in the effluent of the main filter, whereas they were almost stable in the effluents of the bank filtration slow sand filtration combined with underground passage. The elimination rates ranged from 73% to 93% for TCPP, from 71% to 91% for TDCP, and from 80% to 99% for TCEP for the main filter. In the effluent of the bank filtration and slow sand filtration/underground passage, concentrations of TCPP were below LOQ (1 ng L^{-1}) for the whole sampling period, whereas the respective elimination rates ranged from 85% to 94% for TDCP and from 95% to 100% for TCEP.

Figure 2.52 shows the concentrations of chlorinated organophosphates at waterworks B. In comparison to waterworks A, the elimination efficiency of the chlorinated substances by the slow-process sand filtration and underground passage was lower in this water purification plant. Concentrations were reduced from 95 ng L^{-1} to 50 ng L^{-1} (53% elimination) for TCPP, from 37 ng L^{-1} to 14 ng L^{-1} (38% elimination) for TCEP, and from 32 ng L^{-1} to 17 ng L^{-1} (52% elimination) for TDCP. Subsequent ozonization (0.5 g m^{-3} ozone, contact time 0.5 h) did not contribute to the elimination, nor did the multilayer filter consisting of layers of gravel and sand with different grain sizes. After the activated carbon filtration/ UV irradiation, the concentrations of TCPP, TCEP, and TDCP were below LOQ. To examine whether the chlorinated flame-retardants were removed by activated carbon filtration or by UV irradiation, additional samples before and after UV treatment were taken at the same waterworks during a second sampling campaign. The measurements showed that after activated carbon filtration, the concentrations of TCEP, TCPP, and TDCP were below LOQ. Thus, filtration on activated carbon is the most effective treatment step in this waterworks. The measurements of samples from waterworks C confirm the results that ozonization and multilayer filtration did not contribute to the elimination of the chlorinated orga-

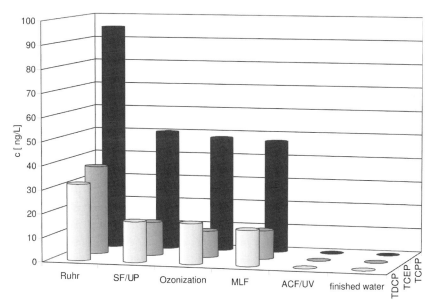

Fig. 2.52 Concentrations (ng L^{-1}) of TCPP, TCEP, and TDCP at different treatment steps at waterworks B (SF/UP: slow sand filter combined with underground passage; MLF: multilayer filter; ACF: activated carbon filter). (Reprint with permission from [129]).

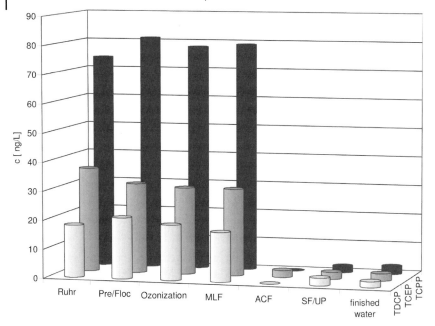

Fig. 2.53 Concentrations (ng L^{-1}) of TCPP, TCEP, and TDCP at different treatment steps at waterworks C (Pre/Floc: precipitation/flocculation; MLF: multilayer filtration; ACF: activated carbon filtration; SF/UP slow-process sand filtration/underground passage).
(Reprint with permission from [129]).

nophosphates. Moreover, TCEP, TDCP, and TCPP were not eliminated by precipitation with aluminum salts and subsequent flocculation, as the concentrations were stable in the raw water and in the effluent of the precipitation. The results for the chlorinated organophosphates are given in Fig. 2.53.

Non-chlorinated Organophosphates
Table 2.34b gives an overview of the concentrations of TiBP, TnBP, TBEP, EHDPP, and TPP in samples taken at waterworks A. Except for TPP, the results were obtained from a five-day sampling period. Data presented for TPP stem from an earlier one-day experiment at the same waterworks. Based on the mean values of the five-day sampling period, for all substances the concentrations measured at the inflow of the gravel pre-filter were similar to those in the inflow of the main filter. This means that the pre-filter did not contribute to the elimination of organophosphate esters in this waterworks. Only for TBEP was a slight reduction in the concentrations observed. In the effluent of the main filter and the bank filtration slow-process sand filtration/underground passage, the measured values of the non-chlorinated alkyl phosphates were below the limit of quantification (LOQ). This means that the biologically active slow-process sand filtration combined with underground passage and underground pas-

Table 2.34b Concentrations of the selected non-chlorinated organophosphates at different treatment steps at waterworks A (PF: pre-filter; MF: main filter; UP: underground passage; MW: mixed water; FW: finished water).

Analyte	Ruhr [ng/L]	Reservoir [ng/L]	PF inflow [ng/L]	MF inflow [ng/L]	MF effluent [ng/L]	UP effluent [ng/L]	MW [ng/L]	FW [ng/L]	Day
	A1	A2	A3	A4	A5	A6	A7	A8	
TnBP	6.5	10	8.4	6.4	6.3	1.9	<1	<1	1
	20	9.1	13	8.8	<1	<1	<1	<1	2
	42	55	34	25	<1	<1	<1	<1	3
	35	28	21	17	<1	<1	3.0	<1	4
	28	35	35	29		<1	1.6		5
Mean	26	27	22	27	1.9	1.0	1.5	<1	
TiBP	36	28	37	40	<3	<3	<3	<3	1
	36	33	34	30	<3	<3	<3	<3	2
	39	32	36	32	<3	<3	<3	<3	3
	43	38	37	34	<3	<3	<3	<3	4
	24	27	26	23		<3	<3		5
Mean	36	32	34	43	<3	<3	<3	<3	
TBEP	170	180	170	130	<3	<3	<3	<3	1
	150	150	150	120	<3	<3	<3	<3	2
	150	160	160	120	3.3	<3	<3	<3	3
	150	150	140	100	<3	<3	<3	<3	4
	140	140	140	99		<3	3.3		5
Mean	150	160	150	110	<3	<3	<3	<3	
EHDPP	0.90	1.0	1.0	0.63	0.17	0.12	0.18	0.15	1
	0.75	0.74	0.74	0.55	0.15	0.17	<0.1	0.14	2
	0.55	0.91	0.65	0.56	<0.1	<0.1	<0.1	<0.1	3
	0.74	0.66	0.55	0.44	<0.1	<0.1	<0.1	<0.1	4
	0.64	0.64	0.67	0.46		<0.1	<0.1		5
Mean	0.72	0.79	0.72	0.53	0.16	<0.1	<0.1	0.15	
TPP		7.2	4.4	3.1	0.30	<0.3	<0.3	<0.3	

sage without additional treatment were effective for the elimination of non-chlorinated organophosphate esters. Moreover, it seems that the elimination efficiency of the main filter concerning the non-chlorinated organophosphates was slightly higher than for the chlorinated substances. In Table 2.34 it is noticeable that the concentrations of TnBP in samples from the Ruhr, the reservoir, and the inflow of the pre-filter varied significantly during the five-day sampling period. No day-to-day variance of the concentrations was observed for TiBP, TBEP, or EHDPP.

At waterworks B the concentrations of the non-chlorinated organophosphates in the raw surface water from the Ruhr (B1) were of the same order of magnitude as those in the raw water from waterworks A. After the water passed the biologically active slow-process sand filter and underground passage (B2), the concentrations of the observed organophosphates T*i*BP, T*n*BP, TBEP, EHDPP, and TPP were below the LOQ. Table 2.35 gives an overview of the results for the non-chlorinated organophosphates in waterworks B.

Figure 2.54 shows the results for TBEP, EHDPP, and the tributyl phosphates in waterworks C. Because of blank values, no data were obtained for TPP. In

Table 2.35 Concentrations of selected non-chlorinated organophosphates at different sampling points in waterworks B.

	T*i*BP [ng/L]	T*n*BP [ng/L]	TBEP [ng/L]	TPP [ng/L]	EHDPP [ng/L]
Ruhr	66	33	140	6.0	1.3
SF/UP	<3	<1	<3	<0.3	<0.1
Ozonization	<3	<1	<3	<0.3	<0.1
MLF	<3	<1	<3	<0.3	<0.1
ACF/UV	<3	<1	<3	<0.3	<0.1
Finished water	<3	<1	<3	<0.3	<0.1

SF/UP: slow sand filtration/underground passage; MLF: multilayer filtration; ACF: activated carbon filtration; UV: UV irradiation.

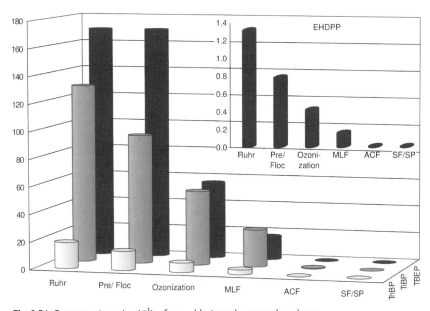

Fig. 2.54 Concentrations (ng L^{-1}) of non-chlorinated organophosphates at different treatment steps at waterworks C (Pre/Floc: precipitation/ flocculation; MLF: multilayer filtration; ACF: activated carbon filtration; SF/SP slow-process sand filtration/underground passage). (Reprint with permission from [129]).

contrast to the chlorinated organophosphates, the concentrations of the non-chlorinated derivates TiBP, TnBP, and EHDPP were reduced by precipitation/flocculation (from 130 ng L^{-1} to 94 ng L^{-1}, elimination rate: 28% for TiBP; from 19 ng L^{-1} to 14 ng L^{-1}, elimination rate: 26% for TnBP; and from 1.3 ng L^{-1} to 0.77 ng L^{-1}, elimination rate: 41% for EHDPP). No effect was observed for TBEP at this treatment step. Moreover, Fig. 2.54 shows that the non-chlorinated organophosphates were eliminated by ozonization (elimination rates between 40% and 67%) and multilayer filtration (elimination rates from 50% to 70%) based on the respective preceding treatment. Although the concentrations were reduced by these processing steps, activated carbon filtration was needed for an effective elimination comparable to that achieved by slow-process sand filtration combined with underground passage in waterworks A.

In contrast to the studies of Stackelberg et al. [122], the selected organophosphates were efficiently removed during drinking water purification. The main differences between the water treatment plant studied by Stackelberg et al. and the waterworks in the Ruhr catchment area are the applied purification techniques. In the U.S. facility, drinking water was purified by addition of powdered activated carbon, flocculation, and filtration through tanks that contained sand and bituminous granular activated carbon (GAC), lignite GAC, or anthracite GAC. In the Ruhr catchment area, on the other hand, drinking water is produced mainly through natural processes or processes that are close to nature, such as bank filtration or groundwater recharge with surface water via slow sand filtration. The natural filter effect of bank zones, soil, and underground passage is supported by biologically active slow-process sand filters and additional preliminary and secondary treatment processes such as precipitation with iron or aluminum salts, ozonization, or activated carbon filtration. In all three waterworks the selected non-chlorinated organophosphates were effectively eliminated by slow-process sand filtration combined with underground passage or bank filtration and slow-process sand filtration/underground passage. The daily variance of the elimination rates for the complete treatment process at waterworks A was low (95 ± 3% for TCPP, 85 ± 8% for TDCP, and 95 ± 5% for TCEP). For the non-chlorinated organophosphates, the concentrations were below the respective LOQ in the finished water. Obviously a very good overview on the elimination efficiency of the selected organophosphates was obtained from this study concerning natural drinking water purification processes at waterworks A, although samples were taken as grab samples. Moreover, the results were not influenced by the daily variance in concentrations that was observed for TnBP and TCEP. Apparently the elimination of the chlorinated substances by means of slow-process sand filtration combined with underground passage depends on the respective conditions, and thus secondary treatment processes such as activated carbon filtration are needed for drinking water purification in some cases. In waterworks A higher elimination rates for TCPP, TCEP, and TDCP (85–95%) were observed for this treatment step compared to waterworks B (38–52%). The difference in elimination efficiency in the respective treatment facilities possibly occurs because of different residence times in the described filters and different soil characteristics. The hydraulic residence time in waterworks A (slow-process sand filtration combined with underground passage) is 10 to 15 days, whereas the contact time in waterworks B is only two to five days. Although the assumed biologically most active area of a slow-process sand filter is supposed to be only the first 3–4 cm and then decreases within the filter bed, it seems that the additional filter effect of the soil is needed for a sufficient elimination of chlorinated organophosphates. The differences between the main filter and the bank filtration combined with slow-process sand filtration and underground passage concerning the elimination of TCPP and TDCP may be a hint as to differences in the biological activity of both treatment processes. The fact that the multilayer filters did not eliminate the selected chlorinated organophosphates can also be traced back to shorter contact times (about 40 min, filter velocity 8 m h^{-1} at waterworks B) in comparison to slow-process sand filtration. Additionally, differences in the biological activity have to be taken into account. Moreover, the non-chlorinated organophosphates were partly eliminated by multilayer filtration at waterworks C, although the elimination efficiency was lower than for slow-process sand filtration/underground passage at waterworks A and B for the same reason. The fact that the multilayer filter in waterworks B did not contribute to the elimination of the chlorinated organophosphates confirms earlier studies in STPs in which similar filters were used for the treatment of treated wastewater before it was discharged to the receiving water [124]. In this case the multilayer filter did not contribute to the elimination of these organophosphates, either. In contrast the previous studies of Stackelberg et al. [122], activated carbon filtration was very effective for the removal of the selected alkyl phosphates. The main differences between the respective filters were on the one hand different contact times and on the other hand differences in the biological activity. Whereas the activated carbon filters in the U.S. facility were biologically inactive and the residence time was only 1.5–3 min, in waterworks B and C contact times were significantly longer (1 h) and the filters were biologically active.

As an alternative to conventional drinking water purification as described in this work, Heberer et al. [123] demonstrated the production of drinking water by applying mobile membrane filter units. These studies revealed elimination rates for TCEP and TCPP of >97.2% and >98.9%, respectively. Although this is a powerful technique, the described units produce comparably small amounts of drinking water ($1.6 \text{ m}^3 \text{ h}^{-1}$ versus $3000 \text{ m}^3 \text{ h}^{-1}$ in the waterworks in this study).

2.4.5.3 Conclusions

Organophosphate ester flame-retardants and plasticizers may be a problem for drinking water production. However, the selected compounds were effectively eliminated in the studied waterworks by slow-process sand filtration, underground passage, and activated carbon filtration. The elimination efficiency of the natural purification processes depends on parameters such as residence time and soil characteristics. For management of activated charcoal filters, it should be taken into account that the sorption capacity of this process is limited and the activated charcoal needs to be replaced on a regular basis to avoid overload.

This study demonstrated that the chlorinated organophosphates TCPP, TCEP, and TDCP were not eliminated by secondary treatment processes such as ozonization or the use of multilayer filters. Because use of these techniques is being discussed for the treatment of treated wastewater to optimize wastewater treatment processes, no effect is expected for chlorinated organophosphates if conditions similar to those in the studied waterworks are chosen. The non-chlorinated derivatives were eliminated by multilayer filtration or ozonization, but the efficiency was lower than for slow-process sand filtration combined with underground passage.

Although organophosphates were detected in surface waters used for drinking water purification, the drinking water quality was not affected by these compounds at the three waterworks in this study. However, it can currently not be excluded that the purified drinking water contains degradation products of the parent compounds. There are plans to investigate this issue further.

2.4.6
Organophosphates and Other Compounds in the North Sea and Lake Ontario: A Comparison

Because the organophosphate flame-retardant TCPP has been found to be non-degradable in wastewater treatment plants and its presence in surface waters was easily proven (see above), the question of whether this compound could reach the marine environment was raised. In a study covering the whole North Sea, a multitude of water samples were subjected to non-target identification with successive tentative quantification. The methods for non-target analysis of organic compounds at ultra-trace levels in seawater used to identify TCPP in the North Sea are described in Section 3.2.2.

It turned out that TCPP was detected in samples from the English coast as well as in the plume of the Rhine and the other German Bight samples. This compound was also determined in samples from the Skagerak (nearly Baltic Sea water) and in Norwegian waters. In Fig. 2.55 mass fragment chromatograms of TCPP in the respective samples compared with a calibration standard are shown. The procedure was tested for blanks, but no blank problems were encountered. No TCPP was found in samples from the central North Sea. An estimation of the concentrations showed that the concentrations ranged from 1 ng L^{-1} to 8 ng L^{-1} in truly marine waters. These concentrations agree with the transport data of the Rhine to the North Sea (compare Section 2.4.3), considering that once each year the water is exchanged totally with water bodies that stem from the central North Sea that are not contaminated with TCPP.

TDCP was detected in a few samples from the German Bight but in none of the samples from other areas of the North Sea in this experiment. It can thus be concluded that:

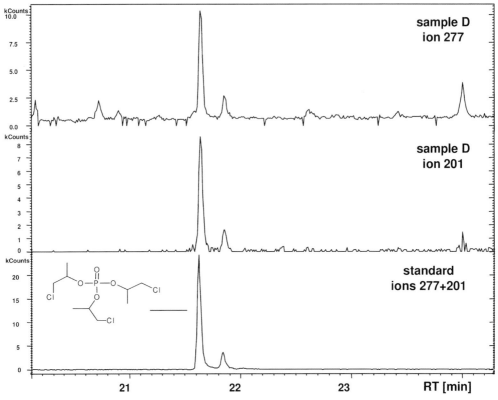

Fig. 2.55 Mass fragment chromatograms of mass fragments 277 and 201 from a sample in the German Bight in comparison to a standard solution. (Reprint with permission from [272]).

1. Usage patterns in Britain and in Germany are different. This assumption is also somewhat supported by manufacturer information.
2. TCPP is very persistent, as it can be detected in waters of the German Bight.
3. TCPP is mobile in the marine environment as well as in limnic systems.

To study these findings in more depth the geographical distribution of organophosphates and other compounds in the German Bight was assessed in comparison to western Lake Ontario (Canada).

Previous studies have shown that organophosphorus flame-retardants and plasticizers are important contaminants in German surface waters, e.g., in the Ruhr and the Rhine. Sewage treatment plants have been identified as point sources of these substances. Because large rivers such as the Rhine and the Elbe flow into the German Bight, it is likely that these compounds will be detected in these marine waters as well. TCPP has already been identified in selected water samples from the North Sea, but the methods were not checked for quantitative results and the measured concentrations were thus only indicative.

In the current study the chlorinated organophosphorus flame-retardants TCPP, TCEP, and TDCP and the non-chlorinated alkyl phosphates TBEP, T*n*BP, and TPP were quantified in the German Bight for the first time. Scott et al. [125] provided a limited dataset on tributyl phosphate and 3-chloroalkyl phosphates in surface waters and precipitation in the Great Lakes region.

Triclosan is also introduced to surface waters via wastewater treatment plants. Only a few datasets are available on surface waters, and no dataset is available on huge water bodies such as Lake Ontario. The presence and emissions of triclosan in wastewater treatment are issues of great concern at the moment. Because no data whatsoever are published on marine or large limnic ecosystems, triclosan and methyl triclosan were included in this study.

Musk compounds are also introduced into surface waters via wastewater treatment. A limited number of datasets on musks in surface waters are available. However, there are only two datasets presenting the discovery of these compounds in large water bodies, e.g., in the North Sea [39] and in Lake Michigan [126], while no published data available on Lake Ontario. An excellent study on HHCB and AHTN in Swiss lakes was performed by Buerge et al. [48]. This study provided the first indications of photolysis of polycyclic musk compounds. However, the concentrations used were milligram per liter, and the experiments were carried out near the surface of the lake; therefore, these experiments do not necessarily give exact data on large, sometimes very turbid ecosystems such as the North Sea or Lake Ontario. None of these studies included the respective transformation product HHCB-lactone, and there have been no attempts to localize the source of pollution and the stability of these compounds within such a huge ecosystem.

In addition to the polycyclic musks, some nitroaromatic musks were analyzed in the study presented in this contribution. It is well known that musk xylene can be transformed to 4-amino musk xylene in the anaerobic parts of wastewater treatment [41, 127, 128]. This compound is supposed to exhibit toxic properties. It has been demonstrated to occur in wastewater treatment plant effluents and in some rivers [44]. However, this transformation product has not yet been studied in huge surface water bodies.

2.4.6.1 Materials and Methods

Chemicals

Tris-(chloro*iso*propyl) phosphate (TCPP) and tris-(1,3-dichloro*iso*propyl) phosphate (TDCP) were obtained from Akzo Nobel (Amersfoort, the Netherlands). These compounds were used without further purification. Technical TCPP gives three peaks at a ratio of $9:3:1$. In this study only the main isomer was used for determination. Tri-*n*-butyl phosphate (T*n*BP), tri-*iso*-butyl phosphate (T*i*BP), triphenyl phosphate (TPP), tris-(2-chloroethyl)-phosphate (TCEP), and tris-butoxyethyl phosphate (TBEP) were purchased from Sigma-Aldrich (Steinheim, Germany). D_{27} T*n*BP was obtained from Ehrenstorfer (Augsburg, Germany). D_{15}

TPP was synthesized from D_6 phenol and phosphoroxytrichloride [129] while AHTN, musk xylene, musk ketone, and triclosan were obtained from Ehrenstorfer (Augsburg, Germany) as pure compound in solution. Methyl triclosan was synthesized from triclosan and trimethyl sulfonium hydroxide [130]. The internal standard D_{15} musk xylene was obtained as a solution from Ehrenstorfer. HHCB-lactone and HHCB were received as a pure standard as a gift from International Flavours and Fragrances (IFF, Hilversum, Netherlands).

All solvents were purchased from Merck (Darmstadt, Germany): acetone (analytical grade/p.a.) and ethyl acetate (analytical grade/p.a. and residue grade/Suprasolv). Toluene and methanol were purchased as residue grade/Suprasolv from Merck. For blank studies and method validation, water (HPLC grade) from Mallinckroth Baker (Griesheim, Germany) was used.

Sampling Area

The German Bight is located in the southern part of the North Sea. It has been considered a region of special concern for fisheries and is probably among the marine ecosystems most heavily influenced by anthropogenic activities such as shipping, fishing, and recreational activities. The German Bight is heavily influenced by the Elbe estuary. The plume of this river extends from the mouth to the north, most of the time east of the island Helgoland. It dilutes into the cleaner waters of the northern North Sea and reaches Danish areas of interest north of the island of Sylt. However, the general direction of the water flow leads to introduction of diluted plume from the Rhine and the Scheld in the west and from the Thames and the Humber in the northwest into the German Bight.

For a comparison Lake Ontario with Hamilton Harbor was chosen, as this is also a large aquatic ecosystem and an area of special concern. The Hamilton Harbor area of concern (AOC) is a 2150-hectare embayment located at the western tip of Lake Ontario and connected to the lake by a shipping canal across the sandbar that forms the bay. Several urban centers, including the cities of Hamilton and Burlington as well as portions of the Regional Municipality of Halton, are located in this watershed.

Sampling

For this study samples from the German Bight were taken during an expedition with the German research vessel *Gauss* from 25 May to 5 June 2005.

The samples were taken with 10-L glass sphere samplers at 5 m below the water surface. For analysis of organophosphate ester flame-retardants and plasticizers, 2 L of each sample were decanted into 2-L glass bottles. All samples were stored at 4 °C until they were extracted with toluene. Sample coordinates are given in Table 2.36.

Samples from Lake Ontario were taken from a small motorboat by submerging capped glass bottles (supplied by the University of Duisburg-Essen laboratory) to a depth of 1 m. Samples were taken on 18 October 2004 (Lake Ontario, samples 1–3) and on 27 October 2004 (samples STP D and 2, sample 4). The respective sampling stations are shown in Fig. 2.56. The samples were taken as

Table 2.36 Sampling position and water salinity of the respective samples from the North Sea.

Sample	Position		Salinity [psu]
	Latitude	Longitude	
1	53°37.2′ N	09°32.5′ E	n.a.
2	53°52.5′ N	08°43.8′ E	15.00 (estimated)
3	54°00.0′ N	08°06.1′ E	32.35
4	54°13.5′ N	08°23.0′ E	28.56
5	54°40.0′ N	07°50.0′ E	31.42
6	55°00.0′ N	08°15.0′ E	29.48
7	55°00.0′ N	07°30.0′ E	32.85
8	54°10.7′ N	07°26.0′ E	33.56
9	54°20.0′ N	06°47.0′ E	33.71
10	54°41.0′ N	06°47.3′ E	34.01
11	54°40.0′ N	06°14.9′ E	34.10
12	54°40.0′ N	05°30.0′ E	34.50
13	54°20.0′ N	05°40.0′ E	34.00
14	53°40.5′ N	06°25.0′ E	32.19

a transect from Hamilton Harbor into Lake Ontario. They were transported to the laboratory by air cargo and extracted on 24 November 2004 immediately after arrival.

Extraction

One liter of the respective samples was extracted with 10 mL toluene after adding an aliquot of internal standard solution containing D_{27} TnBP, D_{15} TPP, and D_{15} musk xylene. The extraction (30 min) was performed by vigorous stirring with a Teflonized magnetic stirring bar. After a sedimentation phase of 20 min, the organic phase was separated from the aqueous one, and the residual water was removed from the organic phase by freezing the samples overnight at $-20\,°C$. The samples were concentrated with a concentration unit (Büchi Syncore, Büchi, Essen, Switzerland) at $60\,°C$ and 60 mbar to 1 mL.

For some of the extracts, it was not possible to remove residual water from the organic phase by freezing. In this case the respective samples were dried over sodium sulfate.

Analytical Methods

The samples were analyzed on a gas chromatography system with mass spectrometric detection (DSQ Thermo Finnigan, Dreieich, Germany) equipped with a PTV injector. The PTV was operated in large-volume injection (LVI) mode (40 μL injection volume) with a sintered glass liner (SGE) with the following temperature program: $115\,°C$ (0.4 min, $130\,mL\,min^{-1}$ He) $\rightarrow 12\,°C\,s^{-1}$ (splitless) $\rightarrow 280\,°C$ (1.2 min) $\rightarrow 1\,°C\,min^{-1} \rightarrow 300\,°C$ (7 min) (cleaning phase).

Sample	Position		Salinity [psu]
	Latitude	Longitude	
1	53°37.2' N	09°32.5' E	n.a.
2	53°52.5' N	08°43.8' E	15.00 (estimated)
3	54°00.0' N	08°06.1' E	32.35
4	54°13.5' N	08°23.0' E	28.56
5	54°40.0' N	07°50.0' E	31.42
6	55°00.0' N	08°15.0' E	29.48
7	55°00.0' N	07°30.0' E	32.85
8	54°10.7' N	07°26.0' E	33.56
9	54°20.0' N	06°47.0' E	33.71
10	54°41.0' N	06°47.3' E	34.01
11	54°40.0' N	06°14.9' E	34.10
12	54°40.0' N	05°30.0' E	34.50
13	54°20.0' N	05°40.0' E	34.00
14	53°40.5' N	06°25.0' E	32.19

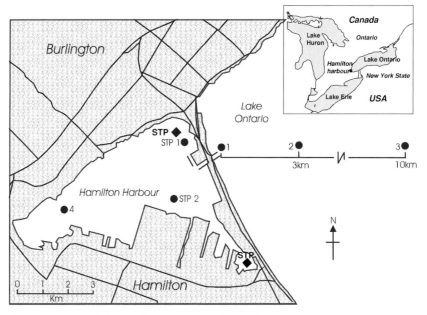

Fig. 2.56 Overview of the sampling area at Hamilton Harbor.
Sampling points: ●; STP effluents: ◆.

The GC separation was performed using a DB5-MS column (J&W Scientific, Folsom, CA, USA) (length: 15 m; i.d. 0.25 mm; film: 0.25 μm) and the following temperature program: $100\,°C$ (2 min) $\rightarrow 30\,°C\ min^{-1} \rightarrow 130\,°C \rightarrow 8\,°C\ min^{-1} \rightarrow 220\,°C \rightarrow 30\,°C\ min^{-1} \rightarrow 280\,°C$ (7 min) using He (5.0) as carrier gas with a flow of

1.3 mL min^{-1}. The mass spectrometer was used with electron impact ionization with 70 eV ionization energy. The MS was operated in selected ion monitoring (SIM) mode. The details of the method, as well as method quality assurance data, are given in Table 2.33.

2.4.6.2 Results and Discussion

North Sea

Chlorinated Organophosphorus Flame-retardants Figure 2.57 shows the distribution of the chlorinated organophosphate esters TCPP, TDCP, and TCEP in the German Bight. The highest amounts were determined at sample station 1 (Elbe estuary near the city of Stade), with TCPP as the dominant substance. The measured concentrations in this sample were 90 ng L^{-1} for TCPP, 22 ng L^{-1} for TCEP, and 15 ng L^{-1} for TDCP. This corresponds with earlier measurements from 2003 (own data). The determined amounts at that time were 160 ng L^{-1} TCPP, 140 ng L^{-1} TCEP, and 10 ng L^{-1} TDCP. Comparable results were obtained for TCPP in 1996 at the same sampling point [131]; concentrations ranged from

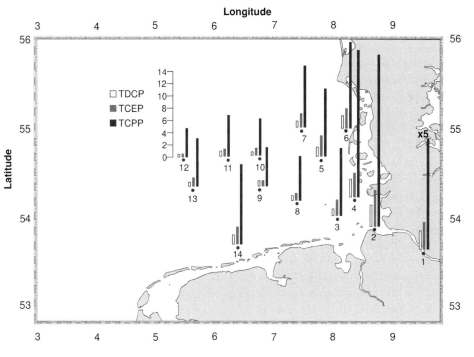

Fig. 2.57 Distribution of chlorinated organophosphate esters in the German Bight of the North Sea; concentrations given in ng L^{-1}.

70 ng L^{-1} to 300 ng L^{-1}. In the region at the mouth of the river (sample point 2), concentrations were noticeably lower (28 ng L^{-1} TCPP, 5.9 ng L^{-1} TCEP, and 3.5 ng L^{-1} TDCP). Sample points 2–7 were located in the plume of the Elbe. Along the coast in a northern direction, the amounts of the chlorinated organophosphates decreased with increasing salinity. Thus, in samples taken at a shorter distance from the coast, the concentrations were higher then offshore. In the Elbe plume amounts detected were approximately 10 ng L^{-1} for TCPP and 1 ng L^{-1} for TDCP and TCPP. Sample points 8–13 were influenced by the inflow of water from the central North Sea that consists mainly of North Atlantic water. Further offshore in a western direction, the concentrations of the selected organophosphates decreased with increasing salinity, indicating a probable dilution with North Sea water. Concentrations were significantly lower than in the Elbe plume and ranged from 4.7–7.2 ng L^{-1} for TCPP and from 0.5–1.0 ng L^{-1} for TDCP and TCEP. Higher amounts of the chlorinated organophosphates were detected at sample point 14. The measured concentrations were 13 ng L^{-1} TCPP, 2.8 ng L^{-1} TCEP, and 1.5 ng L^{-1} TDCP. According to Weigel et al. [272], this sample point is influenced by the plume of the Rhine. The contributing concentrations of the Rhine reach 50–150 ng L^{-1} [121, 132].

To study whether dilution or other processes are relevant for the decreasing concentrations in the more offshore samples, a correlation dilution versus salinity was assessed for these compounds. Equations (3) and (4) were used to correlate dilution factors into pure water from the North Atlantic (salinity 35 psu; practical salinity unit) to normalized concentrations of the respective pollutant.

$$\text{Dilution factor} = \frac{\text{salinity at sampling point}}{\text{salinity of Atlantic water}} \quad (3)$$

$$\text{Normalized concentration} = \frac{\text{concentration at sampling point 1}}{\text{concentration at the respective sampling point}} \quad (4)$$

Figure 2.58 displays the dilution factor (quotient of salinity of Atlantic water and salinity at the respective sampling point) at each sampling point in comparison to the normalized concentrations of TCPP, TDCP, and TCEP under the assumption of an average salinity of 35‰ for Atlantic water. For sample point 2 the salinity was estimated from data achieved from the BSH for the German Bight in 2003 [133]. For the chlorinated organophosphates a linear relationship was observed. This signifies that the decrease in concentrations is mainly attributed to dilution. Moreover, these substances showed a widespread distribution in the German Bight.

Non-chlorinated Organophosphorus Plasticizers and Flame-retardants Similar to the chlorinated organophosphates, the highest amounts of the selected non-chlorinated organophosphates were detected at sample point 1 in the Elbe. The

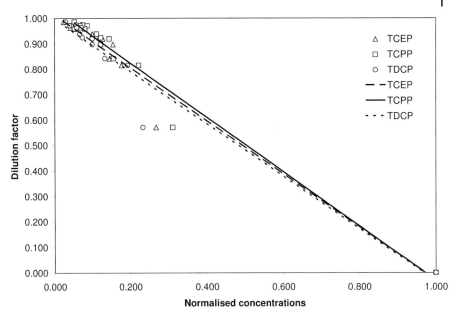

Fig. 2.58 Correlation of dilution factor and normalized concentrations of the selected organophosphorus flame-retardants.

respective concentrations were 23 ng L^{-1} TBEP, 19 ng L^{-1} TnBP, and 3.1 ng L^{-1} TPP. Whereas EHDPP was below LOQ in all samples, it was not possible to determine TiBP because of blank values. The measured amounts correspond with results from samples analyzed in May 2003 (TBEP 24 ng L^{-1}, TnBP 38 ng L^{-1}, and TPP 6.0 ng L^{-1}) near Stade (sample point 1, own data). Figure 2.59 displays the distribution of TnBP, TBEP, and TPP in the German Bight.

In the estuarine region (sample point 2) the concentrations of the selected organophosphates were noticeably lower for TnBP and TPP, whereas the amount for TBEP was below LOQ. In contrast to the chlorinated organophosphates, the concentrations of TnBP and TPP in the Elbe plume were above LOQ only in samples near the coast (sample points 4 and 6). Values for TBEP were below LOQ in all samples, though. Apart from two offshore samples (sample points 10 and 12) in which TPP was detected, the concentrations of all selected non-chlorinated organophosphates were below LOQ. The detection of TPP in the respective samples might stem from the research vessel, as TPP is used, e.g., in hydraulic fluids, or it might be a result of contaminations during the sampling procedure. Because similar starting concentrations of TnBP and TBEP in comparison to the chlorinated organophosphates were detected, a faster reduction in the non-chlorinated compounds was observed. Thus, the conclusion can be drawn that parameters other than dilution with Atlantic water influence the decrease in non-chlorinated organophosphate esters.

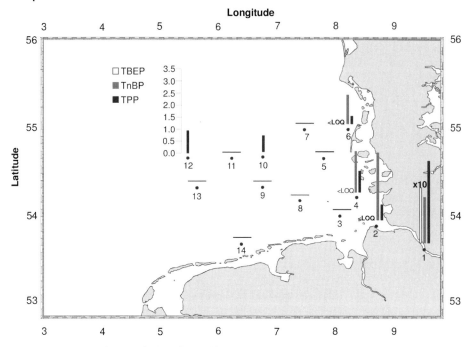

Fig. 2.59 Distribution of selected non-chlorinated organophosphate esters in the German Bight of the North Sea; concentrations given in ng L^{-1}.

Personal Care Compounds   The fragrances HHCB (0.5–5.5 ng L^{-1}), HHCB-lactone (0.2–2.3 ng L^{-1}), AHTN (0.2–1.4 ng L^{-1}), and musk xylene (0.07–0.34 ng L^{-1}) were present in all samples of the German Bight, as shown in Fig. 2.60. Additionally, 4-amino musk xylene was determined at 0.4–5 ng L^{-1} in the same samples.

The lower concentrations of all personal care compounds were generally determined in the offshore samples, while the higher concentrations were found in the more estuarine ones. However, no clear correlation of concentrations could be achieved for the fragrances, e.g., HHCB and AHTN, if all samples were considered. A better correlation was observed for HHCB-lactone. Good correlations (r^2>0.90) were determined for HHCB and AHTN if only those samples from the Elbe estuary were compared with the plume of this river into the North Sea, indicating that some samples, i.e., points 14, 13, 12, 11, 10, and 9, were influenced either by other rivers or possibly by atmospheric deposition and thus should not be included in the correlation. HHBC-lactone is correlated directly to salinity in all samples; thus, a pure dilution of this metabolite is probably taking place (see Fig. 2.61).

There is no indication for de novo formation of this compound from the parent in this marine ecosystem. It is astonishing that the more lipophilic compounds such as musks correlate less with salinity. This indicates relevant inputs into the German Bight other than pure dilution from the Elbe. It might be in-

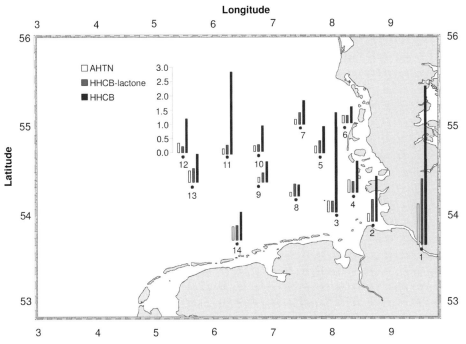

Fig. 2.60 Distribution of the musk fragrances HHCB and AHTN and the metabolite HHCB-lactone in 2005.

teresting to note that the small but significant aberrations are in most cases due to the higher concentrations (i.e., the values in the offshore samples were higher than expected). Thus, it is possible that considerable amounts of musk compounds are transported from the west, e.g., Rhine or Schelde estuary, or northwest, e.g., Thames or Humber estuary, into the German Bight system.

Lake Ontario

Table 2.37 gives an overview of the concentrations of the organophosphate flame-retardants and personal care compounds in Hamilton Harbor and western Lake Ontario. The highest concentrations were found in samples that were influenced by STP outflows (samples STP D and STP E), with highest amounts for TBEP (230–290 ng L^{-1}) and TCPP (69–78 ng L^{-1}). For the other selected organophosphate esters, the respective concentrations ranged from 25 ng L^{-1} for TPP to 49 ng L^{-1} TnBP for the non-chlorinated substances. The chlorinated flame-retardants TDCP and TCEP were in the same range (26–35 ng L^{-1} for TDCP and 35–46 ng L^{-1} for TCEP). At sample point 4 the amounts of all measured organophosphate esters were slightly lower.

Slightly north of the shipping channel (sample point 1), the concentrations of all selected organophosphates were one order of magnitude lower than in the harbor itself (sample point 4). Hamilton Harbor is a more-or-less isolated bay

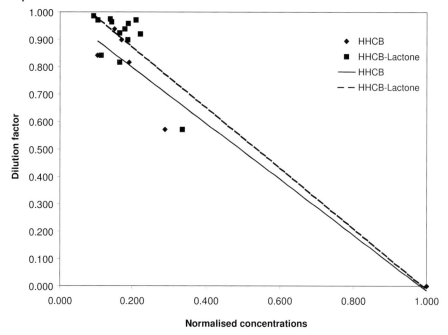

Fig. 2.61 Correlation of dilution factor and normalized concentrations of HHCB and its metabolite HHCB-lactone in the Elbe plume.

with little water exchange with fresh water from Lake Ontario. At a distance of 3 km from the shipping channel (sample point 2), a decrease in the amounts of plasticizers and flame-retardants was observed. These ranged from 0.40 ng L^{-1} TPP to 5.4 ng L^{-1} TBEP for the non-chlorinated substances, whereas they were approximately 3 ng L^{-1} for the chlorinated ones. The reduction in concentrations was most likely a dilution effect from the lake. At a distance of 10 km from the shipping channel (sample point 3), the determined concentrations did not change in comparison to those observed at sample point 2. Thus, no additional dilution was observed. Moreover, this indicates that the different organophosphate esters are stable under the conditions found in Lake Ontario.

A similar pattern was observed for the polycyclic and nitroaromatic musk fragrances as well as for triclosan and methyl triclosan (Table 2.37). The concentrations were high near the STPs (AHTN: 20–28 ng L^{-1}; HHCB: 130–180 ng L^{-1}; HHCB-lactone: 40–70 ng L^{-1}; musk ketone: 0.8–1.8 ng L^{-1}; musk xylene: 0.1–0.3 ng L^{-1}), while the concentrations of triclosan were around 30 ng L^{-1} and its metabolite methyl triclosan ranged around 1 ng L^{-1}. These concentrations were comparable to those found in Germany in waters containing ∼10% treated wastewater [47, 81]. The correlation of HHCB:HHCB-lactone also seems to be similar to the European situations. However, a triclosan:methyl triclosan relationship such as determined in the Hamilton Harbor (20:1) thus far has been

Table 2.37 Overview of the respective concentrations of the diverse emerging pollutants at the sampling points in Hamilton Harbor and Lake Ontario.

Sample	Organophosphates						Personal care compounds						
	TnBP (ng L^{-1})	TPP (ng L^{-1})	TBEP (ng L^{-1})	TCPP (ng L^{-1})	TCEP (ng L^{-1})	TDCP (ng L^{-1})	AHTN (ng L^{-1})	HHCB (ng L^{-1})	HHCB-lactone (ng L^{-1})	Musk ketone (ng L^{-1})	Musk xylene (ng L^{-1})	Triclosan (ng L^{-1})	Triclosan methyl (ng L^{-1})
1	4.6	2.0	18	7.1	5.7	3.7	0.8	7	4	0.04	0.04	1.0	0.1
2	1.6	0.40	5.4	3.5	3.5	2.3	0.2	3	1	0.03	0.07	0.5	0.1
3	1.2	0.34	3.2	3.4	3.5	2.1	0.1	2	1	0.04	0.05	0.4	0.1
4	35	24	170	49	25	19	5.5	41	25	0.12	0.05	11	0.4
STP D	49	25	290	78	46	35	28	180	68	1.8	0.26	29	1.4
STP E	49	26	230	69	35	26	20	130	44	0.84	0.12	27	0.6

determined only for STPs with no denitrification step. This complies with the setup in Hamilton Harbor. Only the Burlington wastewater treatment (STP1) plant is equipped with nitrogen removal, while the one in Hamilton (STP2) is not yet at that stage.

Additionally, the metabolite 4-amino musk xylene was found with a concentration of 1–3 ng L^{-1} within Hamilton Harbor. This compound was not detected in the open water of Lake Ontario. The concentrations in these waters were considerably lower (AHTN: 0.1–0.8 ng L^{-1}; HHCB: 2–7 ng L^{-1}; HHCB-lactone: 1–4 ng L^{-1}; musk ketone: 0.03–0.04 ng L^{-1}; musk xylene 0.04–0.07 ng L^{-1}), while the concentrations of triclosan were around 0.4–1 ng L^{-1} and its metabolite methyl triclosan ranged around 0.1 ng L^{-1}.

The concentrations of the polycyclic musks were of the same order of magnitude as those determined by Peck and Hornbuckle [126] for Lake Michigan (HHCB: 7.9 ng L^{-1}, AHTN: 3.8 ng L^{-1}). The difference between Lake Michigan and Lake Ontario is consistent with population differences, i.e., a larger urban population at the southern end of Lake Michigan compared to western Lake Ontario. However, Lake Ontario also receives upstream inputs from Lake Erie, and the extent to which this contributes to the concentrations of these contaminants in Lake Ontario is not known. For the nitroaromatic musks the differences between Lake Michigan and Lake Ontario are even larger. While in Lake Michigan the concentrations of musk ketone and musk xylene are given as 6.4 ng L^{-1} and 2.8 ng L^{-1}, respectively, the concentrations of these compounds in Lake Ontario are lower by two orders of magnitude. The pattern found in western Lake Ontario is very similar to that determined in European waters. Because consumption and emission patterns of musk ketone and musk xylene in Canada are likely to be the same as in the U.S., the differences between the results from Lake Michigan and this study need further investigation.

The concentrations of triclosan in Hamilton Harbor (11–29 ng L^{-1}) were lower than observed near STPs in the U.S. [134]. They were also lower than those determined in STP effluents, but the values were higher than those determined in riverine or lake water in Europe [80, 81] and they were near to those that were found to have effects on algae [135]. The concentrations of triclosan and methyl triclosan in the open waters of Lake Ontario (0.1–1 ng L^{-1}, Table 2.37) are the lowest concentrations published on these compounds to date. However, especially methyl triclosan is known to accumulate in fish and thus needs to be considered in terms of contamination of fish. Valters et al. [136] have detected triclosan and methyl triclosan in fish in the Detroit River; however, the methyl derivative was only 0.1–0.4% of triclosan in fish plasma.

2.4.6.3 Conclusions

The chlorinated organophosphate esters TCEP, TCPP, and TDCP are persistent organic pollutants that are detected not only in surface waters such as rivers but also in marine water samples. The current study has shown that a decrease in these substances in the German Bight is attributed only to dilution. On the one hand, the determined concentrations of the respective chlorinated organophosphates were only in the lower nanogram per liter range and the bioaccumulation potential is expected to be low due to the log K_{OW} value, but, on the other hand, almost nothing is known about the toxicity of these substances, especially in combination with other synthetic chemicals, although the determined concentrations are lower than effect levels found in laboratory studies. Thus, the widespread distribution of these compounds in the German Bight in addition to the demonstrated persistence in environmental samples has to be regarded as a reason for concern. Apparently, further changes are needed to reach the aim of the Esbjerg declaration: zero emission or background concentrations by 2020 [137]. The objective of this declaration was to ensure a sustainable, sound, and healthy North Sea ecosystem. For that purpose the discharges, emissions, and losses of hazardous substances should be reduced. The guideline principle therefore is the precautionary principle considering zero emissions of synthetic substances into the North Sea.

This study also demonstrated that chlorinated organophosphates exhibit high concentrations compared to other organic pollutants. In the North Sea the concentrations of chlorinated organophosphates are higher than those of the musk fragrances. The same holds true for the samples from Lake Ontario.

The behavior of the selected non-chlorinated organophosphate esters differed to some degree. Whereas the amounts of the chlorinated organophosphates were reduced only by dilution, other parameters might influence the reduction of T*n*BP, TBEP, and TPP as these substances were detected only in samples from the Elbe plume near the coast, even though "starting" concentrations of these substances were in the same range as for the chlorinated alkyl phosphates. The results obtained from the measurements of samples from Lake Ontario confirm the results derived from the German Bight, as a reduction in the concentrations of the selected substances is attributed to dilution as well.

This study has shown that three processes might be relevant for the concentrations of pollutants that enter large, open water bodies such as the German Bight or Lake Ontario:

1. Only dilution is relevant, i.e., chlorinated alkyl phosphates as well as the metabolite HHCB-lactone.
2. Degradation or sorption is important, e.g., the non-halogenated organophosphates are eliminated either by transformation or sorption to suspended matter and successive transport to sediment.
3. Offshore processes such as recycling/remobilization: this is relevant for the more lipophilic compounds such as AHTN and HHCB, which are present at higher concentrations in the offshore areas than are relevant for pure dilution. Either remobilization from sediment or import from western sources seems to be more relevant for these compounds.

2.4.7
Overall Discussion on Chlorinated Organophosphorus Flame-retardants and Other Compounds

Chlorinated organophosphate flame-retardants such as TCPP are omnipresent in limnic waters. They also are present in some coastal marine regions. No data analyzed in this study or other published work show degradation of this compound in the environment. The low elimination rates in rivers and in STPs determined in this study are well in agreement with the data of Kawagoshi [116] as well as with data recently released by the industry in which transformation in closed-bottle tests were reported to start after 28 d, while half-lives of several months were determined [116, 117]. Though its carcinogenicity is debated, with older datasets reporting this compound to the carcinogenic [108] while recent data are indicating that TCPP is not genotoxic and carcinogenic in in vitro tests [138]. Its presence in seawater at nanogram per liter levels makes it a compound of concern or a priority compound under the OSPAR commission regulations. Its presence in surface waters and in some drinking water samples will probably cause challenges under EU chemical regulations as well as under the Water Framework Directive. It would be strongly desirable to find applications that guarantee fewer emissions than the current ones.

2.5
Endocrine-disrupting Agents

2.5.1
Introduction to Endocrine-disrupting Effects

Endocrine-disrupting compounds are chemicals that interfere with the hormonal activities, cycles, or receptors of animals or mankind. Because there are several important hormonal cycles, such as female sexual hormones, male sexual hormones, thyroid hormones, and serotonin hormones, there are also multitudes of enhancers, mimickers, antagonists, etc. It should also be kept in mind that a compound may have different effects in different species or sexes.

In this section we will focus on the estrogen-like compounds. Estrogenic compounds have effects on the fertility of the target organism as well as on non-target species. Also, the parental generation can be affected as well as the following one. A very good overview is given by Colborn et al. [2].

Typical estrogenic compounds are the female sexual hormones such as estrone, estradiol, estriol, etc. However, compounds designed to mimic these hormones to achieve contraception interact with the same acceptors: mestranol, ethinyl estradiol, while diethylstilbestrol (DES) was used in the 1960s to support problematic pregnancies. In the following decades it was shown that DES had effects on the offspring when they became sexually mature.

On the other hand, there are also numerous industrial chemicals (xenobiotics) that interact with the same receptors. Examples include nonylphenol, bisphenol A, phthalates, and pesticides such as DDT. An overview is given in Table 2.38.

It has been hypothesized that for interaction with the respective receptor, a steering group consisting of a polar (e.g., OH) group on a rigid ring system and a bulky, space-demanding alkyl group is preferential. Though this seems to be true in a lot of cases, exemptions such as the case of TBT, which has a completely different molecular structure, are known. TBT is known to have considerable hormonal effects on mollusks, while its effects on other organisms are less well explored.

However, it should be mentioned that some compounds stemming from plants also may interact with the hormonal system, e.g., genistein, coumestrol, and daidzein. The structural formulas are similar. However, the endocrine potency of these compounds is diverse and may differ by several orders of magnitude. In Table 2.38 the diverse effeciencies to the estrone receptor in relation to estradiol are shown.

If these estrogenic compounds show up in the human body they, may have drastic effects on both the male and female body. However, in this book we will focus on effects to the aquatic environment. Male fish especially react to estrogenic compounds, be they natural or synthetic, with the synthesis of egg-yolk protein (vitellogenin); if the concentrations are high enough, actual eggs are produced in the testes. Thus, the virility of wildlife fish is decreased and the population can be maintained only by fish migrating into the affected areas. It has been discussed in the literature that these effects are especially observed in fish populations living near wastewater treatment plant discharges. However, it is not clear at the moment which compounds are the ones relevant for these ef-

Table 2.38 Relative effects of diverse compounds on the estrone receptor implemented into yeast cells (YES assay) [139].

Compound	(EC$_{50}$ Estradiol/ EC$_{50}$ compound)
17 β-Estradiol	1.00
Diethylstilbestrol (DES)	1.57
Coumestrol (phyto)	77
Estriol	273
Dihydrotestosterone	2 000
4-Nonylphenol	5 000
Bisphenol A	15 000
Nafoxidine	34 000
Clomiphene	44 000
ICI 164,384 (standard compound)	64 000
β-Sitosterol (phyto)	220 000
Testosterone	226 000
Methoxychlor (pesticide)	5 000 000
o,p'-DDT (pesticide)	8 000 000

fects in fish. Thus, it is hard to improve the situation before this question is answered. Three scenarios are possible:

1. The naturally occurring hormones may be most relevant, which means that probably only improved wastewater treatment technology would be effective in decreasing the effects.
2. Excretions of oral contraceptives may be the most relevant emissions, which means that altered prescriptions and dosages as well as improved wastewater treatment plant technology would be effective in decreasing the respective effects.
3. The emissions of synthetic chemicals may play the main role in the occurrence of endocrine effects to fish in wildlife. If this is the most relevant scenario, improved handling of chemicals, partial reduction of usage, and improved wastewater treatment plant technologies would be effective in decreasing the respective effects.

At the moment it is hard to decide which of the scenarios is more important, as the concentrations of the respective synthetic chemicals are much higher than those of the naturally occurring hormones. On the other hand, their potency is much lower. Additionally, it should be taken into account that what holds true for one effluent into one river is not necessarily true for another emission into another river. However, at the moment most scientists seem to assume that in most cases the hormones themselves are the most relevant compounds considering aquatic organisms.

When the hormone emissions in wastewater are assessed, it should be noted that not only the parent compounds themselves are excreted but also some conjugated forms, such as sulfates and glucurunates. Additionally, some of the parent compounds can be transformed into each other by relatively simple biochemical processes. Thus, in Table 2.39 some of the major compounds in estrogenic hormone balances are shown. For technical reasons these compounds are analyzed and discussed together with some antibiotics in Table 2.40. Because the analysis of these compounds is by no means easy and straightforward, the analytical chemistry for these compounds is discussed later in detail (see Section 3.1.2).

Table 2.39 Major estrogenic hormones and their conjugation products.

17 β-Estradiol Steroid hormone

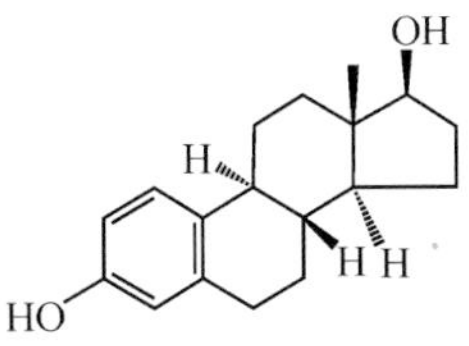

Empirical formula: $C_{18}H_{24}O_2$
MW: 272.18
CAS [50-28-2]
Consumption:
hum: 1.1 t/a [143]
vet: –

Usage: Hormonal treatment for ailments in the climacteric period of women and for prevention of osteoporosis. 17 β-estradiol is a natural steroid hormone that is excreted mainly through the urine of mammals.

Estrone Steroid hormone

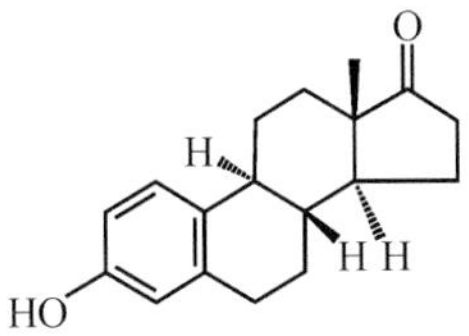

Empirical formula: $C_{18}H_{22}O_2$
MW: 270.16
CAS [53-16-7]
Consumption:
hum: 0 kg/a
vet: –

Sources: Estrone is a metabolite of estradiol that is excreted mainly through the urine of mammals.

Estriol Steroid hormone

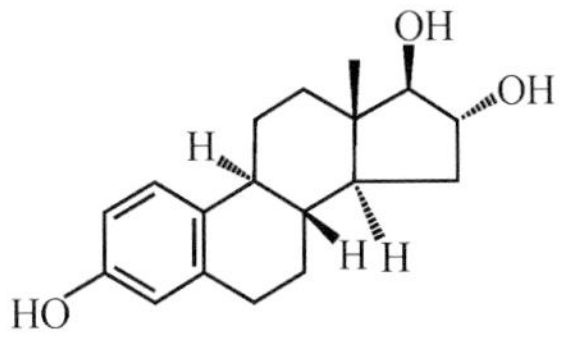

Empirical formula: $C_{18}H_{24}O_3$
MW: 288.17
CAS [50-27-1]
Consumption:
hum: 160 kg/a
vet: –

Usage: Hormonal treatment for ailments in the climacteric period of women and for prevention of osteoporosis. Estriol is a metabolite of estradiol and is excreted mainly through the urine of mammals.

16 α-Hydroxyestrone Steroid hormone

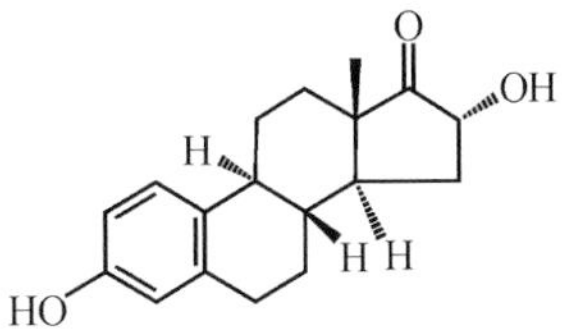

Empirical formula: $C_{18}H_{22}O_3$
MW: 286.16
CAS [566-76-7]
Consumption:
hum: –
vet: –

Sources: 16 α-hydroxyestrone is a metabolite of estradiol that is excreted mainly through the urine of mammals.

Table 2.39 (continued)

β-Estradiol 17-acetate Steroid hormone

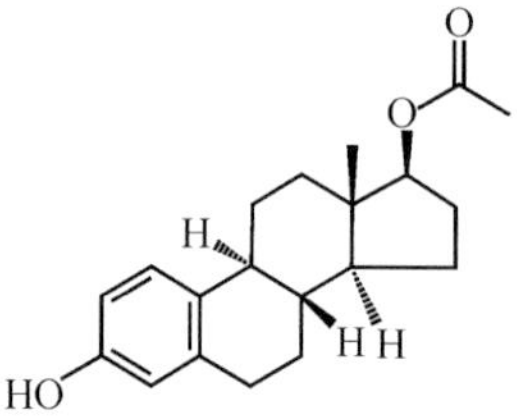

Empirical formula: $C_{20}H_{26}O_3$
MW: 314.19
CAS [1743-60-8]
Consumption:
hum: –
vet: na

Usage: Hormonal treatment for ailments in the climacteric period of women and for prevention of osteoporosis.

β-Estradiol 3-sulfate Hormone conjugate

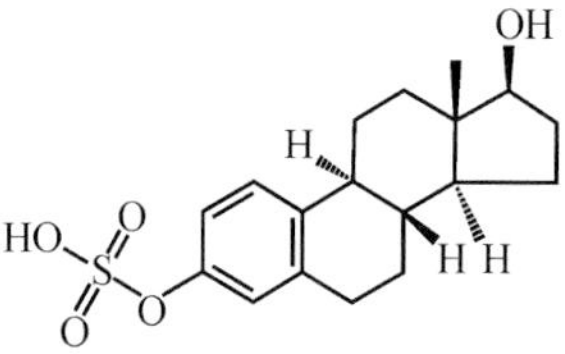

Empirical formula: $C_{18}H_{24}O_5S$
MW: 352.13
CAS [481-96-9]
Consumption:
hum: –
vet: –

Sources: β-estradiol 3-sulfate is a conjugate of estradiol that is formed in the bile and excreted mainly through the urine of mammals.

Estrone 3-sulfate Hormone conjugate

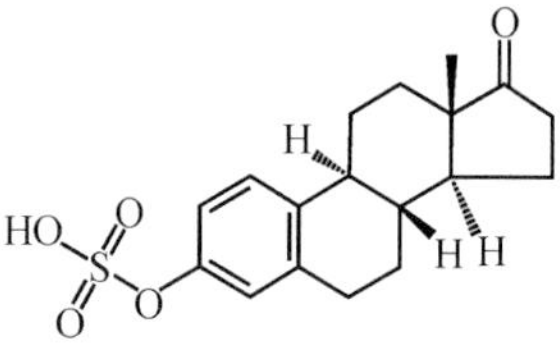

Empirical formula: $C_{18}H_{22}O_5S$
MW: 350.12
CAS [481-97-0]
Consumption:
hum: –
vet: –

Sources: Estrone 3-sulfate is a metabolite of estradiol and a conjugate of estrone that is formed in the bile and excreted mainly through the urine of mammals.

---

17 α-Ethinyl estradiol Oral contraceptive

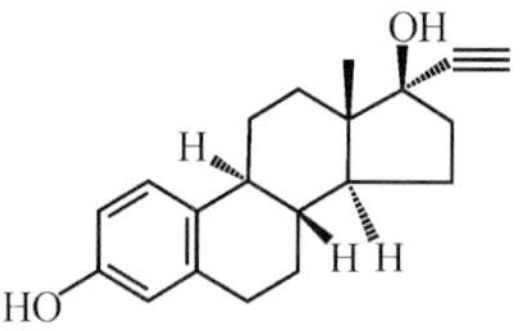

Empirical formula: $C_{20}H_{24}O_2$
MW: 296.18
CAS [57-63-6]
Consumption:
hum: 50 kg/a
vet: n.a.

Usage: Primarily used for oral contraception in human medicine; also used for contraception in veterinary medicine (pet care). Not used in European industrial animal husbandry.

Table 2.39 (continued)

Mestranol Oral contraceptive

Empirical formula: $C_{21}H_{26}O_2$
MW: 310.19
CAS [72-33-3]
Consumption:
hum: 0.8 kg/a
vet: –

Usage: Primarily used for oral contraception in human medicine; also used for contraception in veterinary medicine (pet care). Not used in European industrial animal husbandry.

Testosterone "Male" sexual hormone

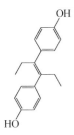

Empirical formula: $C_{19}H_{28}O_2$
MW: 288.43
CAS [58-22-0]
Consumption:
hum: –
vet: –

Usage: No medicinal usage.

Diethylstilbestrol Synthetic compound

Empirical formula: $C_{18}H_{20}O_2$
W: 268.36
CAS [56-53-1]
Consumption:
hum: –
vet: –

Usage: Used to support problematic pregnancies in the 1960s.

Nonylphenol Synthetic compound

Empirical formula: $C_{15}H_{24}O$
MW: 220.18
CAS [84852-15-3]
Consumption: 340000 t/a [149]
hum: –
vet: –

Usage: Monomer of epoxy resins, pesticide formulations, concrete, tensides, etc.

Table 2.39 (continued)

Bisphenol A Synthetic compound

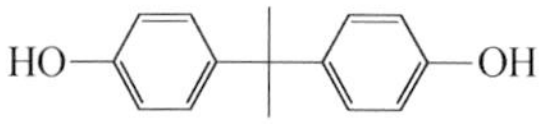

Empirical formula: $C_{15}H_{16}O_2$
MW: 228.29
CAS [80-05-7]
Consumption: 690 000 t/a [140]
hum: –
vet: –

Usage: Monomer of epoxy resins, etc.

Dibutyl phthalate Synthetic compound

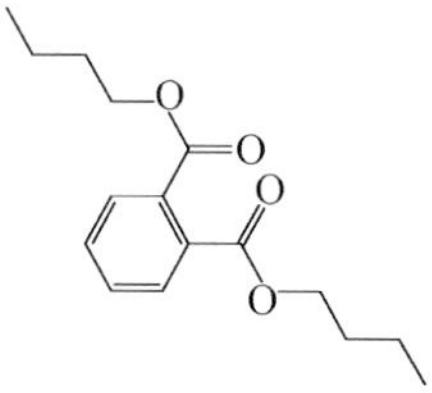

Empirical formula: $C_{16}H_{22}O_4$
MW: 278.35
CAS [84-74-2]
Consumption: 120 000 t/a [141]
hum: –
vet: -

Usage: Plasticizer for PVC and other polymeric resins.

DDT Pesticide (1,1-bis(p-chlorophenyl)-2,2,2-trichloroethane

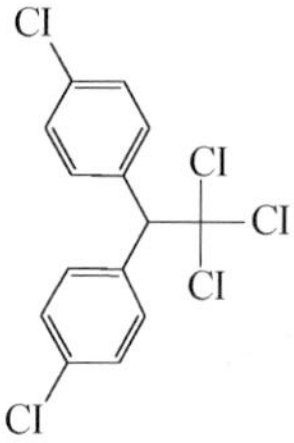

Empirical formula: $C_{14}H_9Cl_5$
MW: 354.49
CAS [50-29-3]
Consumption: developed
countries = 0
hum: –
vet: –

Usage: Pesticide used in malaria prevention.

Tributyltin (TBT) Synthetic biocide

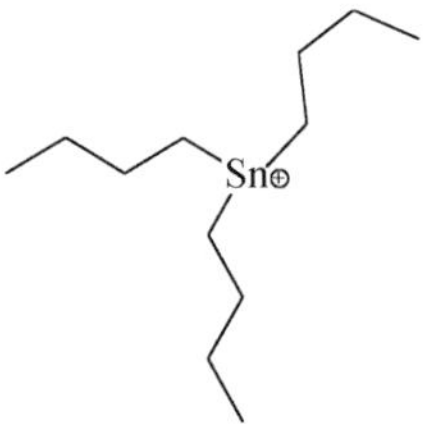

Empirical formula: $C_{16}H_{36}Sn$
MW: 347.17
CAS [1461-25-2]
Consumption: probably several
thousand t/a
hum: –
vet: –

Usage: Biocide used as an antifouling agent in ship coatings; also used as a fungicide in textiles. Has hormonal effects on mollusks.

Table 2.40 Macrolide antibiotics.

Erythromycin Macrolide antibiotic

Empirical formula: $C_{37}H_{67}NO_{13}$
MW: 733.46
CAS [111-07-8]
Consumption:
hum: 19 tons
vet: 20 kg

Usage: Used in human medicine for the treatment of infections such as scarlatina, tonsillitis, erysipelas, pneumonia, diphtheria, pertussis, and acne vulgaris and for the prevention of rheumatic fever. Used in veterinary medicine for the treatment of intestinal infection, mastitis, and pneumonia.

Roxithromycin Macrolide antibiotic

Empirical formula: $C_{41}H_{76}NO_{15}$
MW: 836.52
CAS [80241-83-1]
Consumption:
hum: 9.5 tons
vet: −

Usage: This oxime derivative of erythromycin has a similar application range as erythromycin but is used only in human medicine.

Table 2.40 (continued)

Clarithromycin Macrolide antibiotic

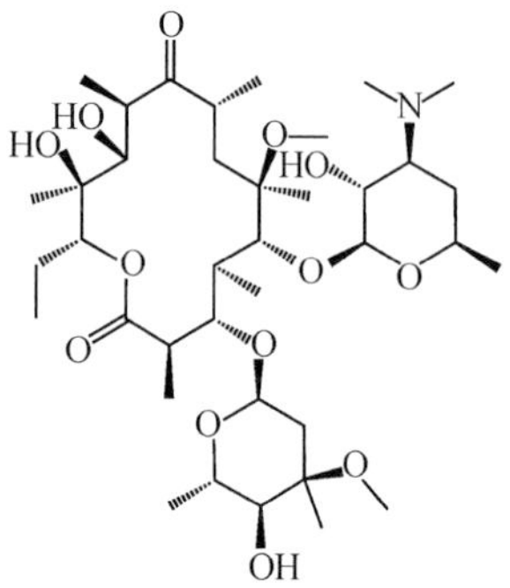

Empirical formula: $C_{38}H_{69}NO_{13}$
MW: 747.48
CAS [81103-11-9]
Consumption:
hum: 7.2 tons
vet: –

Usage: This methyl ether derivative of erythromycin has a similar application range to erythromycin but is used only in human medicine.

2.5.2
Estrogenic Hormones and Antibiotics in Wastewater Treatment Plants

2.5.2.1 Description of the Sample Sites

STP D

Sewage treatment plant D is a middle-sized plant with an inhabitant equivalent value (IEV) of 250 000. The average wastewater inflow per day is 70 000 m³. This plant is equipped with an aerated grit chamber, a primary settling tank, two aeration basins, a circular aeration basin, and a final settling tank. This STP operates with upstream denitrification, while the precipitation of phosphate occurs by means of ferric salts. The inflow was sampled before the processed water of the sludge dewatering was added to the wastewater, and the effluent was tested before the cleaned wastewater was discharged into the river. Figure 2.62 gives more detailed information about the wastewater treatment process of this STP. The samples were taken during the period from 31 August 2004 to 26 September 2004.

STP E

Sewage treatment plant E is smaller than STP D, with an IEV of 64 000. The average wastewater inflow per day is 12 000 m³. This plant is equipped with a neutralization line, an aerated grit chamber, a preliminary settling tank, a trickling filter, a final settling tank, and a poststream-denitrification step. The precipitation of phosphate occurs by means of ferric salts after the trickling filters. The inflow was sampled after the screen cleaner and before the neutralization line, and the effluent was tested directly after the denitrification and before the

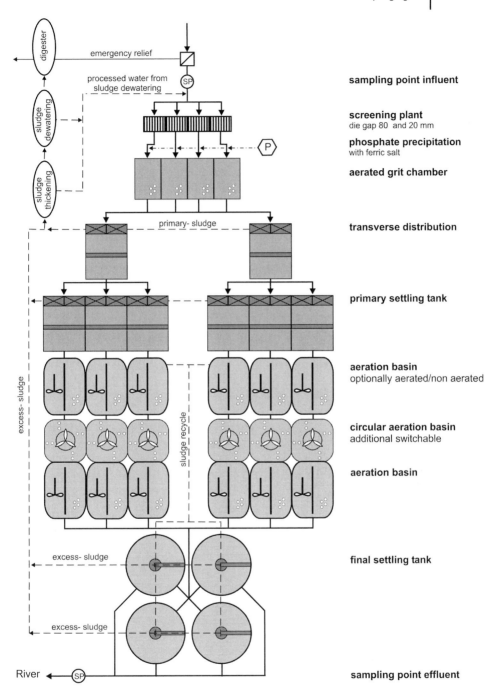

sampling point influent

screening plant
die gap 80 and 20 mm

phosphate precipitation
with ferric salt

aerated grit chamber

transverse distribution

primary settling tank

aeration basin
optionally aerated/non aerated

circular aeration basin
additional switchable

aeration basin

final settling tank

sampling point effluent

Fig. 2.62 Technical sketch of STP D with an IEV of 250000.

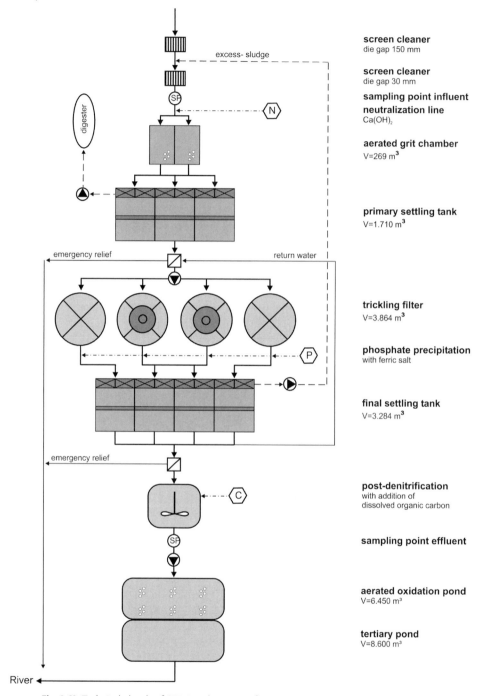

excess- sludge

screen cleaner
die gap 150 mm

screen cleaner
die gap 30 mm

sampling point influent

neutralization line
Ca(OH)₂

aerated grit chamber
V=269 m³

digester

primary settling tank
V=1.710 m³

emergency relief   return water

trickling filter
V=3.864 m³

phosphate precipitation
with ferric salt

final settling tank
V=3.284 m³

emergency relief

post-denitrification
with addition of
dissolved organic carbon

sampling point effluent

aerated oxidation pond
V=6.450 m³

tertiary pond
V=8.600 m³

River ◄

Fig. 2.63 Technical sketch of STP E with an IEV of 64 000.

tertiary ponds. After the tertiary ponds the wastewater is discharged into the river. The technical sketch of this STP provides more insight into this wastewater treatment process (Fig. 2.63). The samples were taken during the period from 28 February 2005 to 30 March 2005.

STP F
Sewage treatment plant F is a small plant with an IEV of 32 000. The average wastewater inflow per day is 13 000 m^3. This plant is equipped with a grit chamber, an aeration basin with simultaneous nitrification and denitrification, and a final settling tank. The precipitation of phosphate occurs by means of aluminum salts at the inflow. The inflow was sampled after the screen cleaner and before the phosphate precipitation, and the effluent was tested before the cleaned wastewater was discharged via final clarifier ponds into the river. Additionally, every three days the effluent of the final settling tanks was also tested.

Detailed information regarding this wastewater treatment process is shown in Fig. 2.64. The samples were taken during the period from 6 June 2005 to 3 July 2005. Supplementary, a 24-h characteristic curve was taken in 2-h steps at the inflow of this plant.

All samples were taken automatically as 24-hour composite samples. The samples were refrigerated at 4 °C during the 24-h intervals. They were transported to the laboratory immediately after sampling and generally were extracted within six hours after arrival. When it was not possible to extract the hormones and antibiotics immediately, the samples were stored at 4 °C for two days maximum. All samples were extracted in duplicates.

2.5.2.2 **Results and Discussion**
During the complete sampling periods of all three STPs, no mestranol, 17a-ethinylestradiol, or β-estradiol 17-acetate could be detected. Detailed information about these procedures will be presented in the final report of the project BASPiK [142].

STP D

Steroid Hormones The daily inflow loads of the single steroid hormones ranged from 0.1 g to 14 g during dry weather conditions, depending on the different types of hormones. During the rainfalls, the wastewater flow rate rose from 40 000 m^3 d^{-1} to 180 000 m^3 d^{-1}. Also, the inflow load of the steroid hormones rose to 20 g d^{-1}. Because the steroid hormones are excreted by humans via urine and because the excretion rate does not rise during rainfalls, the higher loads of these hormones in the wastewater flow may stem from hormones attached to the wastewater canal's sediments, which were transported to the STP due to high flow rates in the sewers.

As examples for all steroid hormones, Figs. 2.65 and 2.66 show the daily loads of estrone and hydroxyestrone during the sampling period.

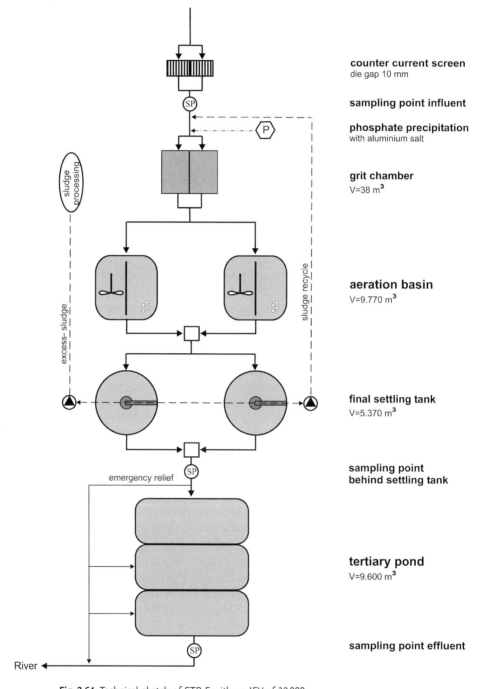

counter current screen
die gap 10 mm

sampling point influent

phosphate precipitation
with aluminium salt

grit chamber
V=38 m³

aeration basin
V=9.770 m³

final settling tank
V=5.370 m³

**sampling point
behind settling tank**

tertiary pond
V=9.600 m³

sampling point effluent

Fig. 2.64 Technical sketch of STP F with an IEV of 32000.

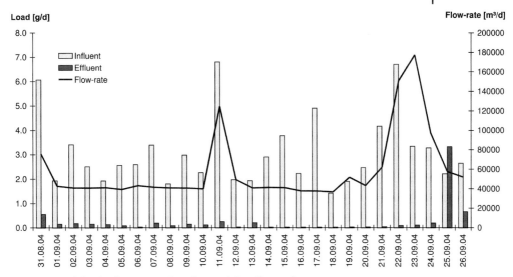

Fig. 2.65 Daily loads of estrone in the inflow and the effluent of STP D over the sampling period. Additionally, the wastewater flow rate during the sampling period is given.

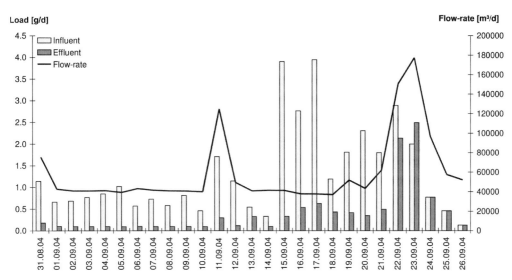

Fig. 2.66 Daily loads of hydroxyestrone in the inflow and the effluent of STP D over the sampling period. Additionally, the wastewater flow rate during the sampling period is given.

The concentrations found in influents were as follows: 19–130 ng L^{-1} for estrone; < LOD–110 ng L^{-1} for hydroxyestrone; up to 68 ng L^{-1} for estradiol; up to 510 ng L^{-1} for estriol; up to 12 ng L^{-1} for estrone 3-sulfate; and < LOD–28 ng L^{-1} for β-estradiol 3-sulfate. The effluents generally had lower concentrations. In the

Table 2.41 Elimination rates of steroid hormones and macrolide antibiotics in the three investigated STPs. Elimination rates are based on the complete mass flow rates during the sampling period. Standard deviations are based on the SD of the validated method and Gaussian error propagation. Negative values indicate "generation"; Positive values indicate "elimination."

	STP D Elimination (%)	STP E Elimination (%)	STP F Elimination (%)
Estrone	92 ± 2	-72 ± 36	50 ± 11
16 β-Hydroxyestrone	69 ± 9	64 ± 11	82 ± 5
17 β-Estradiol	75 ± 3	-53 ± 19	26 ± 9
Estriol	58 ± 14	34 ± 23	69 ± 11
Estrone 3-sulfate	13 ± 16	-360 ± 85	73 ± 5
β-Estradiol 3-sulfate	41 ± 14	-74 ± 15	13 ± 10
Erythromycin	23 ± 16	0 ± 21	15 ± 18
Clarithromycin	-14 ± 10	-3 ± 9	-7 ± 9
Roxithromycin	-19 ± 18	-22 ± 19	-22 ± 27

respective graphs these concentrations are transformed to loads, considering the mass flow of water in the respective time periods.

By comparing the loads of the influents with the load of the effluents, an elimination rate was calculated (Table 2.41). While the elimination of estrone was nearly 100%, the elimination of the other hormones ranged from 41% to 75%, except for β-estradiol 3-sulfate. No significant elimination of this compound could be observed during the wastewater treatment in this plant.

The day-to-day variation of the elimination rate of estrone in this STP is nearly uninfluenced by rainfalls (Fig. 2.67), except that the heavy rainfalls at the end of the sampling period disturbed the elimination of estrone in this STP. As an example for all other steroid hormones, hydroxyestrone showed a decrease in elimination efficiency during rainfalls (Fig. 2.68). At the beginning of the sampling period, the elimination rate of hydroxyestrone was nearly 100%. After the rain event at on 11 September 2004, the elimination rate ranged from 60% to 90%, while no elimination was observed during and after the rain events from 21 September to 26 September 2004.

With regard to the elimination of hydroxyestrone, the biological step of the STP was not able to handle the large inflow of wastewater. Perhaps the bacteria responsible for the elimination of hormones needs time to adapt to the new situation after rainfalls.

Antibiotics The daily inflow load of the single macrolide antibiotics in STP D ranged from 0.9 g to 60 g during dry weather conditions. During the rain events the loads rose to 130 g per day. The concentrations in the influents were 32–1500 ng L^{-1} for erythromycin, 11–60 ng L^{-1} for clarithromycin, and 1.8–155 ng L^{-1} for roxithromycin. The effluents had maximal concentrations of 460 ng L^{-1} (erythromycin), 250 ng L^{-1} (clarithromycin), and 126 ng L^{-1} (roxithromycin).

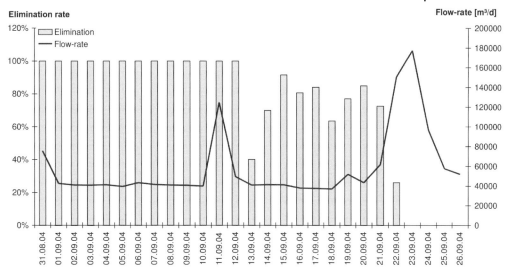

Fig. 2.67 Elimination rates for estrone in STP D during the sampling period in comparison to the wastewater flow rate.

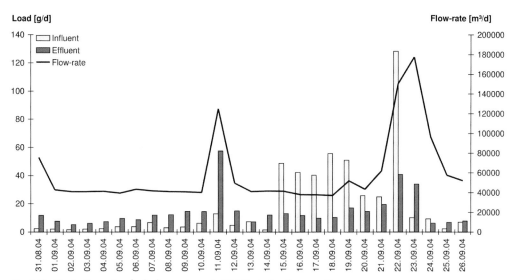

Fig. 2.68 Elimination rates for hydroxyestrone in STP D during the sampling period in comparison to the wastewater flow rate.

As an example for all three macrolide antibiotics, the daily loads of erythromycin during the sampling period are shown in Fig. 2.69. During rain events, the load of antibiotics rose for the same reasons as discussed above for steroid hormones. However, at the beginning of the sampling period, the load of erythromycin in the effluents was higher than in the corresponding influent. This

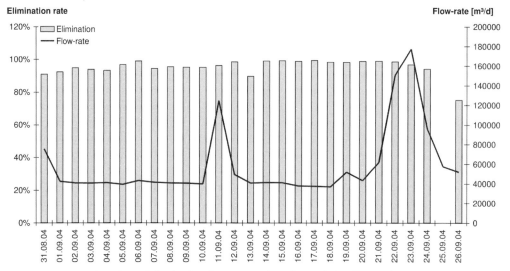

Fig. 2.69 Daily loads of erythromycin in the inflow and the effluent of STP
D over the sampling period.

phenomenon occurred only in this STP. Perhaps consumers of this antibiotic
discharged this drug via the toilet and the active substance was released as a
pulse during the treatment process.

Additionally, the wastewater flow rate during the sampling period is given.
The overall elimination rate of erythromycin in this STP was 23%, while the
semi-synthetic macrolide antibiotics clarithromycin and roxithromycin were not
eliminated. This corresponds to the findings of other authors [143, 144].

STP E

Steroid Hormones The daily inflow load of the respective steroid hormones
ranged from 0.1 g up to 1.5 g during dry weather conditions. For the duration
of the rainfall, the wastewater flow rate rose from 9000 $m^3 d^{-1}$ to 32000 $m^3 d^{-1}$,
while the inflow load of estriol rose to 9 g d^{-1}. The data for estrone and hydroxyestrone are shown in Figs. 2.70 and 2.71, respectively.

By comparing the loads of the inflow with the loads of the effluent, an elimination rate was calculated for each compound (Table 2.41). While the elimination of
hydroxyestrone was 60% (mean value) and that of estriol was 34%, elimination of
the other hormones was not observed. In contrast, an increase of estrone of up to
76% was determined, and the two hormone sulfates rose about 74% and 360%,
respectively. This phenomenon could be explained by the assumption that other
conjugates such as disulfates and sulfate-glucuronides, which were not measured,
were transformed to β-estradiol 3-sulfate during the wastewater treatment. A
transformation of estrone 3-sulfate and other hormone sulfates to β-estradiol
3-sulfate is possible.

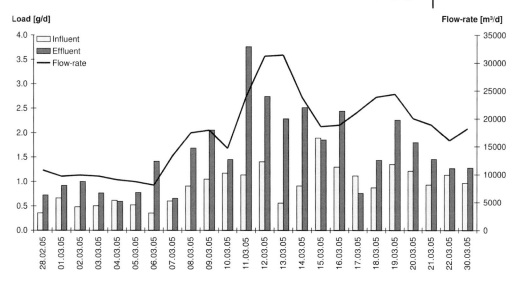

Fig. 2.70 Daily loads of estrone in the inflow and the effluent of STP E (trickling filter) over the sampling period. Additionally, the wastewater flow rate during the sampling period is given.

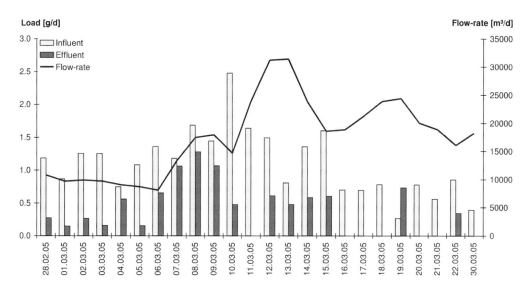

Fig. 2.71 Daily loads of hydroxyestrone in the inflow and the effluent of STP E (trickling filter) over the sampling period. Additionally, the wastewater flow rate during the sampling period is presented.

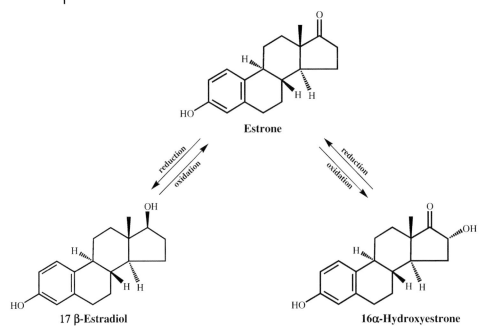

Estrone

17 β-Estradiol

16α-Hydroxyestrone

Fig. 2.72 Possible transformation route of hydroxyestrone to estrone and estradiol in STP E.

The increase in steroid hormones based on the transformation of hydroxyestrone to estrone and estradiol is shown in Fig. 2.72. This transformation is described in the literature [145]. The sum of the total inflow load of these three hormones in STP E was 54 g, while the sum of the total outflow load was 53 g. Thus, no elimination of steroid hormones could be achieved by using a trickling filter for wastewater treatment.

The elimination of 16 α-hydroxyestrone as well as estriol during the complete sampling period also showed a decrease in the elimination efficiency during rainfall. This STP type is also vulnerable to heavy rain events (Fig. 2.73). It can thus be assumed that in this case hydroxyestrone was mainly transformed.

Antibiotics The daily inflow load of the respective macrolide antibiotics ranged from 0.3 g to 7 g. The concentrations in influents were 73–650 ng L^{-1} for erythromycin, 97–690 ng L^{-1} for clarithromycin, and 16–250 ng L^{-1} for roxithromycin. The effluents had maximal concentrations of 440 ng L^{-1} of erythromycin, 480 ng L^{-1} of clarithromycin, and 350 ng L^{-1} of roxithromycin.

As an example for all three macrolide antibiotics, Fig. 2.74 shows the daily loads of erythromycin during the sampling period.

As shown in Fig. 2.74, the macrolide antibiotics were not eliminated in this STP during the sampling period. Thus, the trickling filter technique does not eliminate macrolide antibiotics during wastewater treatment.

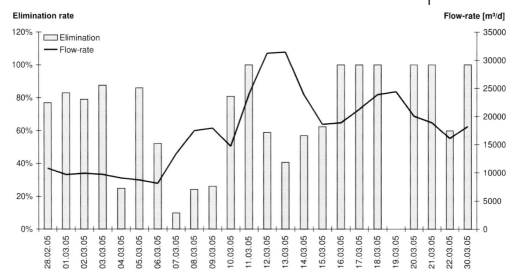

Fig. 2.73 Elimination rates of hydroxyestrone in STP E (trickling filter) during the sampling period in comparison to the wastewater flow rate.

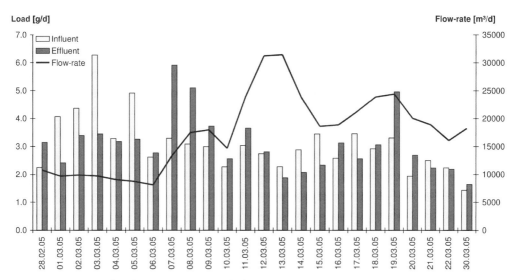

Fig. 2.74 Daily loads of erythromycin as an example for the macrolide antibiotics in the inflow and the effluent of STP E (trickling filter) over the sampling period. Additionally, the wastewater flow rate during the sampling period is presented.

STP F

Steroid Hormones Figure 2.75 shows a diurnal cycle of the inflow loads of estrone, hydroxyestrone, and estriol. Samples were taken in two-hour intervals

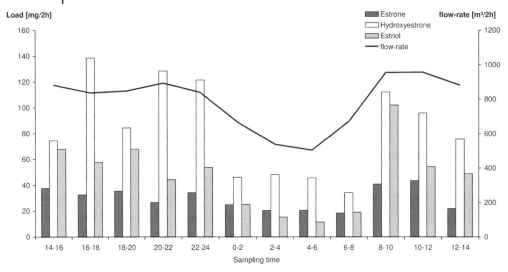

Fig. 2.75 Two-hour inflow loads of three hormones over one day in STP F.

over a period of 24 hours. The loads decreased during nighttime and rose in the morning and evening. While the flow rate decreased about 50% during the night, 16 α-hydroxyestrone decreased from 140 mg in the hours between 16:00 and 18:00 to 38 mg in the hours between 6:00 and 8:00. All other steroid hormones behaved similarly. The exhibited overnight decrease and morning increase indicates that the steroid hormones stem from excretion by humans. Because most people sleep during the night, the excretion of hormones via the urine is low. Due to the morning act of urination, the excretion of the hormones increases.

A 24-h flow-controlled composite sampling was important for this project because one-time sampling of the influent and effluent gives only a snapshot with no information about elimination rates.

The daily inflow load of the respective steroid hormones in STP F ranged from 0.1 g to 4.2 g during dry weather conditions. During rainfalls, the wastewater flow rate rose from 8000 m^3 d^{-1} to 35 000 m^3 d^{-1}. The inflow loads of the respective steroid hormones also rose during this period. The effluent loads of the final settling tanks were in the same range as the effluent loads of the tertiary ponds. Generally, these effluent loads are shifted because the tertiary ponds hold a volume of 9600 m^3, which is a hydraulic retention time of one day. However, this means that this final clarifier, which was a simple concrete basin with little biological activity, did not contribute to the elimination of these compounds.

As an example for all steroid hormones, the daily loads of estrone and hydroxyestrone in STP F during the sampling period are presented in Figs. 2.76 and 2.77. In the influent samples, the following concentrations were found: 14–87 ng L^{-1} for estrone, 15–190 ng L^{-1} for hydroxyestrone, up to 18 ng L^{-1} for estradiol, up to 440 ng L^{-1} for estriol, up to 26 ng L^{-1} for estrone 3-sulfate, and

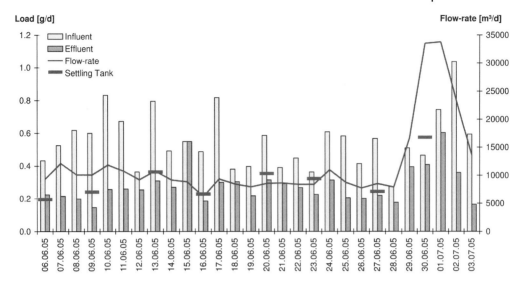

Fig. 2.76 Daily loads of estrone in the inflow and the effluent of STP F as well as the effluent of the settling tank over the sampling period. Additionally, the wastewater flow rate during the sampling period is presented.

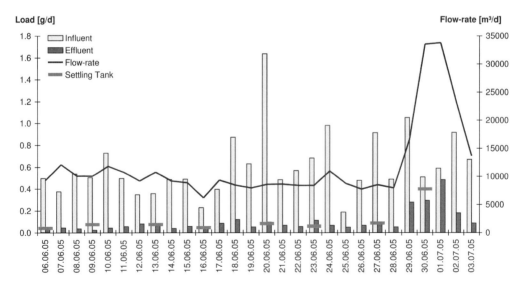

Fig. 2.77 Daily loads of hydroxyestrone in the inflow and the effluent of STP F as well as the effluent of the settling tank over the sampling period. Additionally, the wastewater flow rate during the sampling period is presented.

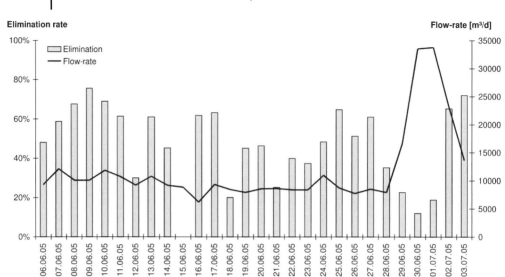

Fig. 2.78 Elimination rates of estrone in STP F during the sampling period in comparison to the wastewater flow rate.

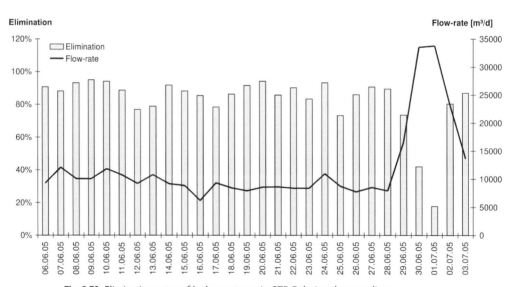

Fig. 2.79 Elimination rates of hydroxyestrone in STP F during the sampling period in comparison to the wastewater flow rate.

< LOD–28 ng L^{-1} for β-estradiol 3-sulfate. The concentrations in the effluent samples were generally lower than those found in the influent samples. However, the data for β-estradiol 3-sulfate revealed a different fate for this compound.

By comparing the loads of the influents with the load of the effluents, an elimination rate was calculated (Table 2.41). The elimination of the hormones ranged from 30% to 82%, except for β-estradiol 3-sulfate. This hormone conjugate showed no significant elimination during wastewater treatment. This could be explained by the assumption that other conjugates such as disulfates and sulfate-glucuronides, which were not measured, were transformed to β-estradiol 3-sulfate during the wastewater treatment. A transformation of estrone 3-sulfate and other hormone sulfates to β-estradiol 3-sulfate is possible as well.

The elimination rates of estrone (Fig. 2.78) and hydroxyestrone (Fig. 2.79) as examples for all hormones, except for β-estradiol 3-sulfate, show a dependency on the flow rate. The sampling in the first 22 days was during a dry weather period. The elimination rates varied extremely from 20% to 80% for estrone, while they were nearly constant for hydroxyestrone (75% to 95%). Perhaps the STP was too small for a continuous stable elimination of the steroid hormones.

Antibiotics The 24-h characteristic curve of the antibiotics given in Fig. 2.80 shows a dependency on the time of the day. While the flow rate decreased about 50% during the night, erythromycin decreased from 1.3 g in the hours between 20:00 and 22:00 to 0.2 g in the hours between 6:00 and 8:00. All other macrolide antibiotics behaved similarly. This overnight decrease and morning increase indicates that macrolide antibiotic concentrations stem from excretion by humans. The well-described "pee peak" is also observable. Additionally, a 24-h flow-con-

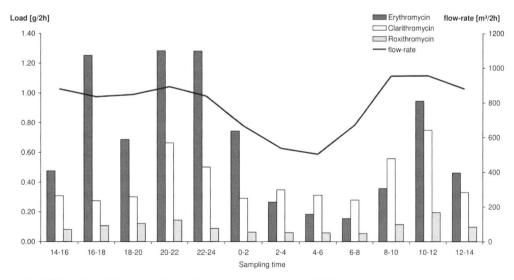

Fig. 2.80 Two-hour inflow load of three hormones over one day in STP F.

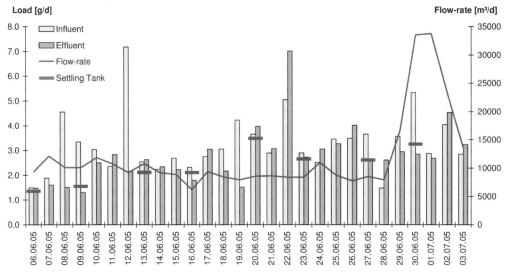

Fig. 2.81 Daily loads of erythromycin as an example for the macrolide antibiotics in the inflow and the effluent of STP F as well as the effluent of the settling tank over the sampling period. Additionally, the wastewater flow rate during the sampling period is presented.

trolled composite sampling is recommended for the determination of elimination rates for antibiotics.

The daily inflow loads of the respective macrolide antibiotics ranged from 0.4 g to 14 g, while the concentrations in influents were 85–780 ng L^{-1} for erythromycin, 92–1500 ng L^{-1} for clarithromycin, and 25–94 ng L^{-1} for roxithromycin. The effluents had maximal concentration of 830 ng L^{-1} for erythromycin, 520 ng L^{-1} for clarithromycin, and 170 ng L^{-1} for roxithromycin. As an example for all three macrolide antibiotics, the daily loads of erythromycin during the sampling period are shown in Fig. 2.81. The macrolide antibiotics were not eliminated significantly in this STP during the sampling period. Thus, a smaller dimension of STPs with a simultaneous denitrification step does not help to eliminate macrolide antibiotics during wastewater treatment.

Comparison of STP D, STP E, and STP F
When all investigated STPs are compared, the differences between STP E and STP D are especially remarkable. STP E is equipped with a trickling filter and has an IEV of only 64 000, while the more modern STP D has an IEV of 250 000 (Table 2.42). This comparison of loads shows that the amount of the respective steroid hormones released into the environment per day is up to a factor of 6 higher in the smaller STP E than in the activated sludge plant (STP D).

No significant elimination of the macrolide antibiotics could be detected in any of the three STPs investigated in this study. The daily discharge of these compounds rose as expected with the number of affiliated persons (Table 2.42).

Table 2.42 Comparison of the daily discharge of steroid hormones and macrolide antibiotics in the three investigated STPs.

	STP D (IEV 250000)	STP E (IEV 64000)	STP F (IEV 32000)
	Discharge (mg d^{-1})	Discharge (mg d^{-1})	Discharge (mg d^{-1})
Estrone	250	1600	280
16 α-Hydroxyestrone	410	390	110
17β-Estradiol	340	290	84
Estriol	2100	3200	360
Estrone 3-sulfate	270	760	28
β-Estradiol 3-sulfate	510	390	270
Erythromycin	15000	3100	2800
Clarithromycin	8800	4400	3700
Roxithromycin	4600	1900	1300

2.5.2.3 Conclusions

It has been demonstrated that steroid hormones and macrolide antibiotics are released into the environment via the pathways of humans, urine, wastewater, STPs, and STP effluents. Three different types of STPs were investigated to determine their elimination rates.

Larger STPs eliminated hormones more constantly than small STPs. Heavy rainfall events, which resulted in high wastewater flow rates, led to a collapse of the biological treatment concerning these compounds. More rain storage basins are necessary to reduce these influences on the wastewater treatment process.

The steroid hormones could not be eliminated by means of the trickling filter technique during wastewater treatment. Only a transformation of the hormones among each other was observed. This technique should be replaced with more state-of-the-art treatment techniques.

No significant elimination of macrolide antibiotics could be detected in the three STPs. New concepts of treatment should be developed for the elimination of macrolide antibiotics during wastewater treatment if environmental issues are being taken into consideration.

2.5.3
Nonylphenol and Other Compounds in the North Sea

Endocrine-disrupting agents are among the most discussed in environmental chemistry as well as from the administrators' side. In several fish populations male fish have started to feminize, e.g., by production of the egg-yolk protein vitellogenin or even by producing eggs in male gonad tissue [3]. While performing experiments for identification/quantification, it was thus decided to include nonylphenols into the studies, especially as it became obvious that these compounds are indeed introduced to the sea via the estuaries. Nonylphenols were not studied in detail in riverine ecosystems or STPs, as there is already a multitude of published data on these compounds.

Nonylphenols (NPs) are known to exhibit endocrine-disrupting properties [146–148]. While there are just a few applications of nonylphenol itself, nonylphenol polyethoxylates (NPEOs) are used as detergents. Though NPEOs are no longer used as surfactants in domestic (household) applications, they are still used as industrial detergents in metal and textile processing [148]. Additionally, applications in the paper industry, in production of paints, and in formulations of pesticides are relevant [146]. Ahel et al. [149] reported the use of 340 000 t NPOEs worldwide after the banning of NP derivatives in domestic detergents. These large amounts are discharged directly or via STPs [27] into rivers. In STPs or rivers, a transformation of NPOEs to NPs takes place, e.g., via decarboxylation processes [146, 150]. In chemical terms, NPs are a complex mixture of diverse isomers, all containing branched alkyl groups with a sum formula of C_9H_{19}. This variety of isomers makes analytical procedures quite difficult.

In addition to their endocrine-disrupting properties NPs are considered to be toxic to fish and daphnia [151]. These compounds have been detected in a variety of large rivers (e.g., [55, 152]) and in a few coastal waters [153–157]. Marcomini et al. [158] even tried to establish balances of NPEOs and linear alkylbenzene sulfonates (LAS) in the lagoon of Venice. Nonylphenols have only once been observed in open seawater, i.e., in the Sea of Japan [159]. Thus, the following questions were raised: Are NPs and other endocrine-disrupting agents present in the North Sea, and do they show effects in the marine environment? These issues are of special importance because regulating bodies such as the Oslo and Paris Commissions (OSPAR) consider endocrine-disrupting agents to be of high priority. This study was performed to gain insight into the question of whether these compounds are present in marine ecosystems.

Identification and quantification of NPs were performed for the first time in open North Sea water up to 100 km from the coast. During these experiments the presence of bis(4-chlorophenyl)-sulfone in the water samples also was detected. Because this compound has been detected in fish only once [162], but never in seawater and very rarely [160, 161] in fresh waters, it was considered worthwhile to study this compound as well. Little is known about the ecological and toxicological properties of this compound. It seems to be used as a plasticizer in polymers made for high-temperature applications [162]. Thus, this paper presents for the first time the presence and the concentrations of bis(4-chlorophenyl)-sulfone in seawater. Additionally, linear alkylbenzene sulfonates (LAS) were detected in estuarine waters during a project on identification of organic pollutants in the marine environment.

The combination of these compounds had its origin in an identification/ quantification study in sediment and water samples of the German Bight. These compounds are considered by the author to be reliably identified, besides those that have already been published [39, 60, 62, 64, 163–167].

2.5.3.1 Materials and Methods

Seawater

Nonylphenols were analyzed by the method for lipophilic compounds in seawater high-volume extraction (see Section 3.2.1). In brief, nonylphenol was analyzed in selected ion monitoring (SIM) mode of the mass spectrometer at 107.050 amu, 200 ms dwell time and 135.081 amu 200 ms dwell time, with 130.992 amu (of perfluorotributylamine) lock mass at 50 ms dwell time and a cycle time of 1.115 s. Eight chromatographic peaks were obtained from the technical mixture on mass 107.050, while six were detected on mass 135.081. Three of those were identical to those obtained on the other mass. All these peaks were integrated and the results of the samples were compared with the signals obtained from a standard solution (taking the internal standard into account). The final results were calculated as a sum of all isomers. For quantification and calibration, a mixture of NP isomers from the technical mixture obtained from Riedel de Haen (Seelze, Germany) was used, as no pure isomers were available. To check the extraction procedure, several spiked waters were extracted. Water samples obtained from a Seral (Ransbach, Germany) apparatus were spiked with acetonic solutions of a multicomponent standard solution to reach nanogram per liter concentrations. (The final acetone concentration was below 1%.) The recoveries as well as the limit of detection are shown in Table 2.43. The LOQ was determined during a multipoint calibration with standard solutions, and the signal-to-noise values were obtained. The solution that gave a s/n <10 was defined to be below the limit of determination (quantification). This concentration was transferred to the extract concentration and thus to the water concentration by taking the respective enrichment (concentration) factors into account.

This result was verified with real samples and blank values. The recovery rates were not taken into account for calculating the final concentrations, because these were performed as a qualitative test with only three concentrations rather than a thorough recovery study as is usually done in our laboratory for quantitative work.

Bis(4-chlorophenyl)-sulfone was analyzed at 158.967 amu, 150 ms dwell time and 285.962 amu, 150 ms dwell time. Mass 263.987 (of perfluorotributylamine) was used (with 50 ms dwell time) as lock mass. The pure compound was obtained from Merck (Darmstadt, Germany) and was used for calibration. To check the extraction procedure, several spiked waters were extracted. The recoveries as well as the limit of detection are shown in Table 2.44. The recovery rates were not taken into account for calculating the final concentrations; thus, the concentrations given are lower than the real ones.

Table 2.43 Limit of quantification (LOQ), recovery rate (rr), and blank (obtained from an empty sampler) of NPs from seawater.

	LOQ (ng L^{-1})	rr (%)	Blank (ng L^{-1})
Nonylphenols	0.01	41	<0.01

Table 2.44 Limit of quantification (LOQ), recovery rate (rr), and blank (obtained from an empty sampler) of bis(4-chlorophenyl)-sulfone from seawater.

	LOQ (ng L^{-1})	rr (%)	Blank (ng L^{-1})
Bis(4-chloro-phenyl)-sulfone	0.10 [a]	56	0.052

a)   Due to blank problems.

Details of the method applied in this study are documented in Section 3.2.1. In general, nonylphenol-polyethoxylates $(NP(EO)_n; n > 2)$ and LAS were extracted from 4-L seawater samples with Carbopack solid-phase extraction (SPE) cartridges. The cartridges were dried in a stream of nitrogen and eluted with 2 mL methanol. NPEOs and LAS were determined by tandem mass spectrometry (MS/MS) after RP-HPLC separations. Defined mixtures of NPEO and LAS were used for calibrations and standard additions. For coupling the HPLC system with the mass spectrometer, the conditions for APCI ionization using ammonium acetate were chosen as follows: vaporizer temperature, 400 °C; capillary temperature, 180 °C. Corona voltage was operated at 5 kV. The potential of capillary, tube lens, and API octapole was chosen as 50, 50, and –3 V, respectively. Sheath gas pressure was operated at 40 psi. The electron multiplier was operated at 1200 V and the conversion dynode at 15 kV. In MS-MS mode the ion source pressure was set to 0.5 Torr. Under collision-induced dissociation (CID) conditions, the pressure in quadrupole 2 (collision cell) was adjusted to 1.3 mTorr. For NPEOs the atmospheric pressure chemical ionization (APCI) parent ion scan m/z 291 in the positive mode was applied to quantify the area under the peaks. LAS were quantified in the same way, applying the negative parent ion mode (parents of m/z 183 amu) after electrospray ionization (ESI) using a TSQ 700 mass spectrometer (Finnigan MAT, San Jose, USA). LC/MS analyses were performed, recording ESI mass spectra scanning from 100 to 1200 u at 3 s. For HPLC-MS/MS analyses in the parent ion mode, the scan ranges were 300–1200 amu (parents 291 amu) or 200–400 amu (parents 183 amu), respectively. Separations of the SPE eluates of water samples containing LAS and nonylphenol polyethoxylates $(NP(EO)_n; n > 2)$ were operated for qualitative and quantitative determination on a Nucleosil C_{18} (5 μm, spherical) column (25 cm×4.6 mm i.d.) (CS, Langerwehe, Germany). The mobile phase was acetonitrile (HPLC grade) from Promochem and Milli-Q-purified water. The flow rate for column separation using a gradient elution (solvent A: acetonitrile; solvent B: water) was 1.0 mL min^{-1}, and the gradient was programmed as follows. Starting with 10% A: 90% B, the concentration was increased linearly to 90% A within 15 min and was held for 15 min. After passing the inline UV diode array detector (Waters 996 DAD system in combination with a Millenium 2010 data system), 0.5 mL min^{-1} of 0.1 M ammonium acetate was added, which resulted in an overall flow rate of 1.5 mL min^{-1}. The limit of detection was 10 ng L^{-1} and 5 ng L^{-1}, respectively, for NPEO and LAS, and the analyses of NPEO and LAS were performed with 6% and 3% RSD, respectively.

Sediment Samples for Nonylphenol-polyethoxylates and LAS

Nonylphenol-polyethoxylates (NP(EO)$_n$; $n > 2$) and LAS were extracted from freeze-dried sediment samples by Soxhlet extraction using methanol. After evaporation to dryness, the samples were reconstituted in 200 mL of Milli-Q-purified water applying ultrasonification. NPEOs and LAS were extracted using solid-phase extraction (SPE) cartridges filled with RP-C$_{18}$ material from Baker (Deventer, Netherlands). For this purpose water samples were forced through the SPE cartridges after passage through a glass-fiber filter. After elution with 2 mL methanol, the samples were analyzed by HPLC-MS/MS without any further cleanup. Chromatographic conditions for NPEO and LAS determination by MS/MS in the parent ion mode were applied as described before.

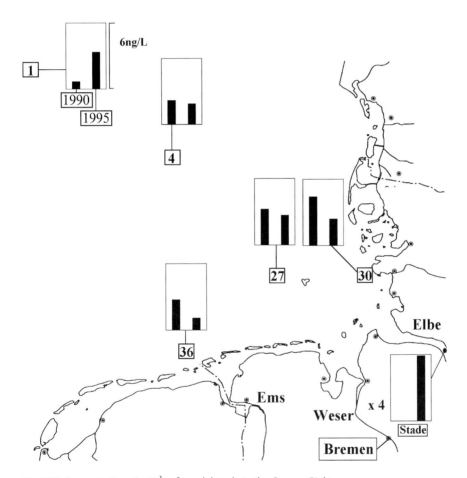

Fig. 2.82 Concentrations (ng L^{-1}) of nonylphenols in the German Bight. Values for 1990 and 1995 are given for comparison. (Reprint with permission from [168]).

2.5.3.2 **Results**

Seawater

Nonylphenols/nonylphenol-polyethoxylates Retention times (RTs) and spectra of nonylphenols were identical to the technical standard, though the intensity pattern of the signal in the gas chromatogram varied in some cases from the used isomeric mixture. SIM measurements were performed in seawater extracts. In some samples the difference in pattern was significant, indicating some transformation processes.

Quantifications were performed from the respective SIM measurements. The results ranged from 33 ng L^{-1} in the estuary to less than 1 ng L^{-1} in the central North Sea (see Fig. 2.82). The concentrations decrease significantly with distance from the coast. Though a high variation between 1990 and 1995 was determined, no significant trends in concentration in the years 1990 and 1995 were obvious. NPEOs could not be detected in the water samples from the estuary (LOQ: 10 ng L^{-1}).

Bis(4-chlorophenyl)-sulfone The presence of bis(4-chlorophenyl)-sulfone was identified by the mass spectrum (Fig. 2.83) obtained from a water sample from 1990. The base peak was analyzed with an exact mass measurement. The ob-

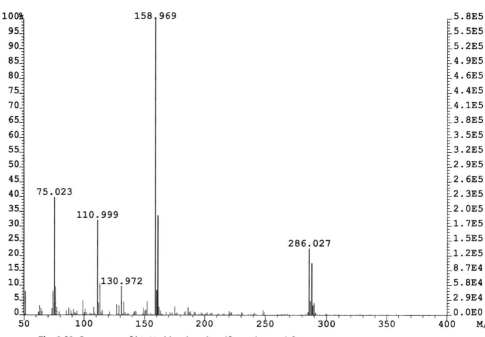

Fig. 2.83 Spectrum of bis(4-chlorphenyl)-sulfone obtained from a water sample from the Elbe estuary in 1995. (Reprint with permission from [168]).

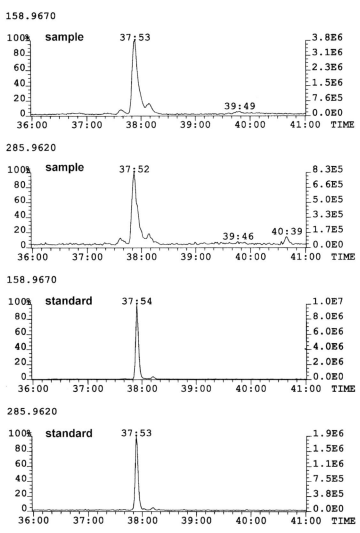

Fig. 2.84 Mass fragment chromatograms (measured in SIM mode) of masses 158.967 and 285.962 significant for bis(4-chlorophenyl)-sulfone. The upper chromatograms originate from a sample westward of the island Helgoland taken in 1990, while the lower ones originate from a standard solution. (Reprint with permission from [168]).

served mass of 158.969 amu for the base peak is in good agreement with the theoretical value of C_6H_4ClOS (M-C_6H_4ClO) that was calculated as 158.967 amu (compare Fig. 2.83). The hypothesized presence of this compound was verified by injecting a pure standard solution in comparison with a sample. Retention times and spectral data were found to be identical (Fig. 2.84).

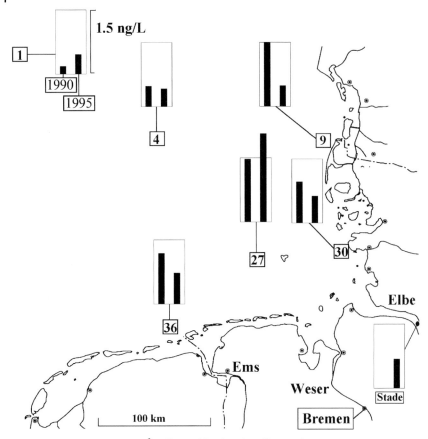

Fig. 2.85 Concentrations (ng L^{-1}) of bis(4-chlorphenyl)-sulfone in the German Bight. Values for 1990 and 1995 are given for comparison. Maximal values within the fields were 1.5 ng L^{-1}. (Reprint with permission from [168]).

The concentration range for bis(4-chlorophenyl)-sulfone was found to vary from 0.18 ng L^{-1} to 2.2 ng L^{-1} (Fig. 2.85) in the seawater. The highest concentration was found near the island of Helgoland, while the concentration in the Elbe estuary was 0.68 ng L^{-1}. No changes in concentrations in the seawater considering the years 1990 and 1995 were determined.

LAS were analyzed in several samples from the Elbe estuary, and all samples gave positive results. The maximum concentration found was 30 ng L^{-1}.

Sediments

Nonylphenol-polyethoxylates (NP(EO)$_n$; n > 2) and LAS   Samples of marinas were taken in 1997, and four selected samples were used for examination. Accumulated NPEO concentrations of compounds with more than 2 EO units varied from <10 ng g^{-1} up to 39 ng g^{-1} dry matter, whereas concentrations of LAS ranged from 39 ng g^{-1} to 106 ng g^{-1} dry matter.

Nonylphenols and Nonylphenol-monoethoxylates (NP1EO)   The data from Bester et al. [168] are cited for comparison. Briefly, samples of Wadden Sea sediments and North Sea sediments were taken in 1997. The concentrations of NP ranged from 10 ng g^{-1} to 153 ng g^{-1} in dry matter, with the highest values found in Husum and the lowest in the Baltic Sea marina Kappeln. Regarding the open-sea samples, only sample 13 (obtained 30 km from the island of Helgoland) gave significant concentrations of nonylphenols. No NP1EO was detected in any of the obtained samples. This finding is probably due to the fact that NP1EO is less hydrophobic than NP, and thus this compound may be diluted in the water phase or possibly biodegraded to nonylphenol.

2.5.3.3 Discussion

Nonylphenols
A dilution profile from the Elbe estuary to the central German Bight is quite obvious. Station 36 was selected to check for Rhine River contribution, as the water in the German Bight flows eastward along the coast. From Fig. 2.82 it can be concluded that there is an input from the Rhine into the North Sea as well. A tentative comparison between salinity and nonylphenol concentrations for the year 1990 showed a linear correlation, thus indicating no strong sedimentation or biodegradation pattern, although nonylphenols have been found in sediment samples at concentrations 1000 times higher than those in the water. Another study showed that nonylphenols could in principle be degraded even in seawater and marine sediments if higher concentrations than those observed in this study are applied [169]. A comparison of the correlation of nonylphenols versus salinity and of lindane (γ HCH) versus salinity shows a similar situation for both compounds (data discussed in more depth in Section 4.3; the data for HCH were taken from Koopman et al. [170]). The slopes are nearly identical, which means that the basic behavior of both compounds in the marine environment is also very similar. HCH isomers are also believed to be mostly solved in the aqueous phase and not strongly sorbed to particulate matter or sediment. Additionally, the biodegradation of HCH isomers in seawater is probably very low. The difference in concentrations during the degradation studies (ppb) and the real environment (ppt) may provide an explanation for these different results [169]. Degradation often occurs at high concentration, while at lower concentrations no degradation can be observed. It may be concluded from these data that both HCHs and NP are of similar persistence in the marine environment.

   In a comparison of the years 1990 and 1995, there was no indication of a trend in time, despite the huge political and economical changes in the drainage area of the Elbe River.

Effects of nonylphenols to oyster embryos have been demonstrated by Nice et al. [171]. The lowest concentration used in this experiment was 100 ng L^{-1}. This concentration still had significant effects on malformations, and it is of the same order of magnitude as that found in the water of the Elbe estuary. Billinghurst et al. [172] also found effects on barnacle settlement starting at concentrations of 100 ng L^{-1}. The same group published data on effects on barnacles at concentrations of 10 ng L^{-1} as well [173], which is just slightly higher than the concentrations observed in North Sea waters. The combination of these two results strongly indicates possible effects to marine wildlife, especially in the estuaries.

Station 30 has already been used for comparisons of diverse compounds [163]. In the current study we found that the concentrations of nonylphenols (2.5 ng L^{-1} at station 30 in 1990) were relative high in comparison with other priority pollutants such as PCBs and were of the same order of magnitude as the alpha isomer of hexachlorocyclohexane (a HCH), which is used as a kind of marker in marine environmental chemistry. A comparison of concentrations as well as a discussion on diverse marine pollutants is given in Section 4.3. The concentrations found in the Sea of Japan by Kannan et al. [159] ranged from 0.002 ng L^{-1} to 0.093 ng L^{-1}. In comparison to those concentrations, the ones in the North Sea are about three orders of magnitude higher. This is quite astonishing because most of Kannan's samples were taken quite close to the Japanese mainland. Additionally, it is surprising that these authors found the highest concentrations at a depth of 1000 m. It is possible that the pollution near Japan is quickly diluted by the ocean currents of the deep sea around Japan, as indicated by the authors. This is not possible in the rather shallow waters of the North Sea, which is about 30 m deep at the sampled area.

NPEO in Seawater

The presence of nonylphenol-polyethoxylates (NP(EO)$_n$; n>2) in seawater could not be confirmed in these experiments because of their easy degradability under aerobic conditions. NPEO derivatives are carboxylated either in the alkyl and polyether chain or in both positions [174], whereas NPEOs as precursors of the nonylphenols were found in marina sediments in concentrations varying from <10 ng g^{-1} to 39 ng g^{-1} dry matter. These compounds were found because of widespread industrial applications and their discharge with the wastewater into the aquatic environment. They were found, e.g., in the waters of the Elbe and the Saale [175, 176] at concentrations ranging from 0.2 µg L^{-1} to 18 µg L^{-1}.

LAS in Seawater

LAS were found at concentrations of up to 30 ng L^{-1} in seawater samples, whereas Gonzalez-Mazo et al. [177] determined LAS in the Bay of Cadiz (Spain) at concentrations three orders of magnitude higher (range of <10–200 µg L^{-1}). The LAS found in the marina sediment in our study probably originated from the discharges of municipal wastewater treatment plants.

Bis(4-chlorophenyl)-sulfone in Seawater

The distribution of bis(4-chlorophenyl)-sulfone is quite peculiar. It is rather evenly distributed, with elevated concentrations near the coast (stations 36, 30, 27, and 9). This leads to the assumption that this compound has been introduced from nonpoint sources for quite some time. Because the concentrations in the Elbe were quite low in 1995, it seems that this river was not the dominant source of bis(4-chlorophenyl)-sulfone in that year. It is a matter of speculation whether the source in the river decreased or whether other sources were more dominant. The concentration of bis(4-chlorophenyl)-sulfone in comparison is rather low. However, considering that there is little knowledge about the toxicity of this compound and that it seems to be quite stable in the marine environment, it is a potential threat to the marine ecosystem, especially because this compound is bioaccumulating in fish [178].

2.5.3.4 **Conclusions**

- Endocrine-disrupting agents such as nonylphenols and NPEO are present in the water of the German Bight at concentrations near those for which effects have been described.
- Nonylphenols are present in sediment samples from Wadden Sea marinas.
- Nonylphenols are transported at least several hundred kilometers into the sea.
- Because these compounds are obviously persistent, it should be studied whether effects from this compound are to be expected.
- It needs to be discussed whether these compounds should be treated as priority pollutants with regard to the Oslo and Paris Commissions for the protection of the North Sea and the Atlantic, respectively. Recently, nonylphenol was included on the priority list of the Water Framework Directive [4].
- More organochlorine compounds such as bis(4-chlorophenyl)-sulfone are present in the water of the German Bight.
- LAS has proved to be a not-easily-degradable surfactant and is present in seawater and several marine sediment samples.

Currently, the concentrations of NP in STP effluents are still in the microgram per liter range; thus, concentrations in the environment are still high despite the voluntary restriction (Freiwilliger Selbstverzicht) on the usage of NP and NPEO in 1990. STPs do eliminate these compounds reasonably well, but the elimination power of sewage treatment plants is not sufficient to reduce the discharged amounts to lower than $\sim 5\ \mu g\ L^{-1}$. However, in 2004 a legal restriction on nonylphenol became effective. It will be interesting to see whether the concentrations of these compounds decrease in aquatic environments in the near future [179, 180].

2.6
Diverse Compounds

2.6.1
Benzothiazoles in Marine Ecosystems

Benzothiazoles were included in the marine studies because as they may serve as an indicator for rubber wear-off, which is also a type of lifestyle indicator as it is closely connected to automobile traffic. Thiocyanatomethylthiobenzothiazole is used as biocide and has a field of application similar to that of triclosan.

In three previous studies, methylthiobenzothiazole was identified in the water of some estuaries [153, 181, 182]. However, concentrations of this compound were determined in only one study [181]. In addition, methylthiobenzothiazole (MTB) and benzothiazole (BT) (see Fig. 2.86) were detected in marine water during an ecotoxicological experiment by Bester et al. [183] at Helgoland in the central German Bight as well as in two European rivers [55, 184].

Because quantitative information about these compounds in the marine environment was lacking, it was considered essential to study the spatial distribution of these compounds in the water of the German Bight of the North Sea. MTB as well as other benzothiazoles are used for vulcanization of rubber tires. In California these compounds have been used as a tracer for street runoff water [181].

Another derivative of benzothiazole, i.e., thiocyanatomethylthiobenzothiazole (TCMTBT), is used as a fungicide in leather processing [185], wood protection [186], and ship and boat antifouling paints [187]; TCMTBT can be transformed to MTB and BT [188, 189] as well as to mercaptobenzothiazole, all of which proved to be comparatively stable in a study in aquatic and terrestrial ecosystems [188]. Accordingly, this biocide was also included in the present study. Toxic and ecotoxic effects of derivatives of benzothiazoles have been described by Reemtsma et al. [188] concerning bacteria (*Vibrio fischeri*) and inhibition of nitrification, while Hendriks et al. [184] provided toxicity data on *Daphnia magna*. To the authors' knowledge, food chain processes have not been studied so far.

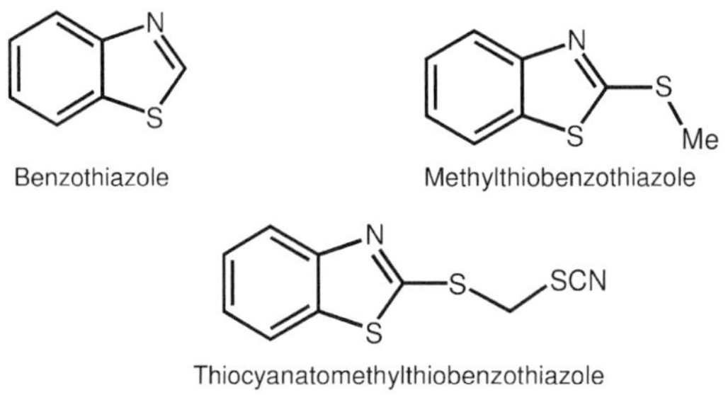

Benzothiazole   Methylthiobenzothiazole

Thiocyanatomethylthiobenzothiazole

Fig. 2.86 Structural formulas of benzothiazole, methylthiobenzothiazole, and thiocyanatomethylthiobenzothiazole.

It was the aim of this study to identify and quantify BT, MTB, and TCMTBT in water samples from the German Bight of the North Sea, as at least some of them are highly bioactive [188, 190]. To the authors' knowledge, the presence and concentrations of these compounds have never before been reported for seawater.

2.6.1.1 Materials and Methods

The identification/quantification procedure is described in depth in Section 3.2.1. For quantification purposes, the samples were cleaned up with a 1-mL silica solid-phase extraction cartridge with the eluent n-hexane:dichloromethane 10:1 for the first fraction and dichloromethane:*iso*propanol 1:1 for the second one. No further fractionation was necessary. Blanks were measured by extracting a sampler during the experiment with 1 L n-pentane. The respective extract was processed the same way as the water extracts used for quantification.

For quantification the same gas chromatographic conditions were used as before (Sections 2.1.4 and 3.2), while the mass spectrometer was operated in the selected ion monitoring mode with resolution m/Äm = 1000, ion source 200 °C, 70 eV EI. Benzothiazole was quantified by measuring ions 135.014 amu (atomic mass units) and 108.003 amu with a dwell time of 200 ms and lock mass 130.997 amu of PFTBA (perfluorotributylamine) (50 ms). The cycle time within this function was 1.032 s. For MTB the ions 181.002 amu and 108.003 amu were analyzed with 400 ms dwell time each, using the same lock mass and a cycle time of 1.002 s. TCMTBT was analyzed at 179.994 amu, 350 ms dwell time with 130.992 amu lock mass at 50 ms and a cycle time of 1.115 s.

The standard compound benzothiazole (CAS No. 95-16-9) was purchased from Merck (Darmstadt, Germany), while methylthiobenzothiazole (MTB) (CAS No. 64036-43-7) was obtained from Aldrich (Steinheim, Germany). Thiocyanato-methylthiobenzothiazole (CAS No. 21564-17-0) was re-crystallized from the raw product.

2.6.1.2 Results

Methylthiobenzothiazole (MTB) was identified within this study by using the mass spectrum for library research. This result was confirmed by comparing the respective mass spectra and retention times with those obtained from injecting a pure standard solution. The mass of the molecular ion measured at resolution 500 was 180.046 amu and was in agreement with the theoretical mass of the molecular ion of MTB ($C_8H_7NS_2$) 181.002 amu within the precision of this measurement. The agreement is even better for M – Me (measured 166.005 amu versus 165.976 amu [theory for $C_7H_4NS_2$] and M – SH (measured 148.043 amu versus 148.022 amu [theory for C_8H_6NS]). Because both the high-resolution mass spectra and the retention times of MTB in the sample and the standard solution were identical, the identification of this compound in the water sample from the Elbe and the German Bight was considered to be confirmed (see Fig. 2.87 a).

The presence of benzothiazole in water samples from the Elbe estuary was confirmed by comparing mass spectra and retention times to those obtained from an original standard (see Fig. 2.87 b). The molecular ion of 135.078 amu measured at resolution 500 agrees with the theoretical value of C_7H_5NS (135.014 amu). However, quantification data for BT were handled with care because a blank occurred for this compound, although the blank concentrations of benzothiazole are considerably lower than those in almost all samples (Table 2.45).

TCMTBT can be analyzed by gas chromatography, though the response is comparatively low. The quotient area per picogram injected is one order of magnitude higher for benzothiazole and MTB compared to TCMTBT.

Recovery rates were analyzed by spiking 1 L water (8–800 ng L^{-1} benzothiazole and 4-400 ng L^{-1} MTB, respectively) with the respective standard solution in acetone (200 µL) and extracting this sample with 10 mL *n*-pentane, i.e., the recovery rates refer to the extraction part of the procedure only. Therefore, the recovery rates were not used for correcting the respective concentrations discussed below. The recovery rates, detection limits (LOQ), and blanks for the respective compounds are documented in Table 2.45.

Table 2.45 Limits of quantification (LOQ) and blanks obtained by 100-L samples. Recovery rates (rr) were measured by extraction of 1-L samples and given in virtual water concentrations.

	LOQ (ng L^{-1})	rr	Blank (ng L^{-1})
Benzothiazole	0.005	0.22	0.18
MTB	0.01	0.54	<LOQ
TCMTBT	0.3	0.35	<LOQ

During June 1990 and from June to July 1995, several 100-L water samples (unfiltered) were taken in the German Bight within the national German monitoring program (BLMP) in order to quantify the concentrations of diverse "classical" pollutants such as HCHs, PCBs, and PAHs. These samples were also used for analyzing the concentrations of MTB, benzothiazole, and the fungicide TCMTBT. The latter could not be detected in the samples, which may be due to the low detection limits, while benzothiazole and MTB were present in all estuarine and marine water samples investigated. Ion fragment chromatograms (measured in SIM mode) for MTB and benzothiazole obtained from a water sample and a standard solution, respectively, are shown in Figs. 2.88a and 2.88b. Unfortunately, no quantification is available for the estuarine sample from 1990, as this was used for identification of the compounds investigated herein.

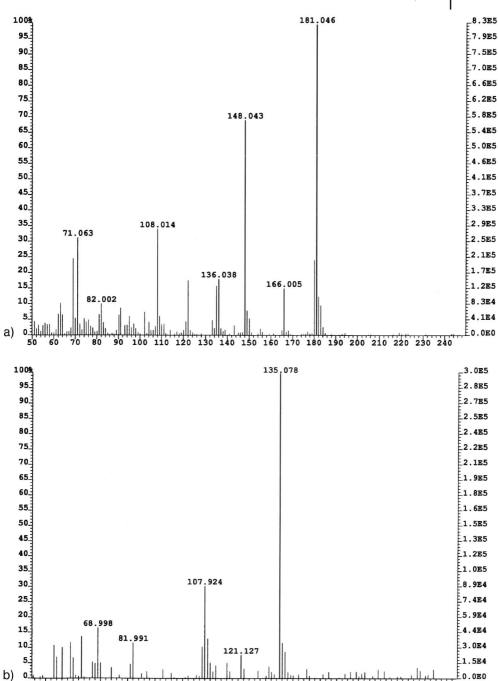

Fig. 2.87 (a) Mass spectrum of methylthiobenzothiazole obtained from a water sample from the Elbe estuary. (b) Mass spectrum of benzothiazole obtained from a water sample from the Elbe estuary. (Reprint with permission from [60]).

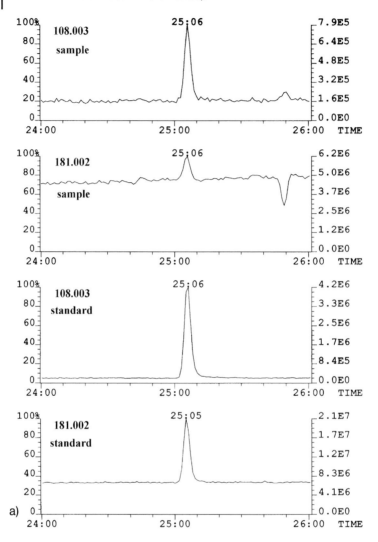

Fig. 2.88 (a) Ion fragment chromatograms (in SIM mode) of masses 108.003 and 181.002, significant for MTB from station 9 taken near the island of Sylt (upper two chromatograms) and a standard (lower two chromatograms). (Reprint with permission from [60]).

The observed concentrations are shown in Fig. 2.89 and Table 2.46. The amounts of MTB found in the Elbe estuary near Stade in 1995 were quite high (55 ng L^{-1}), while they were about three orders of magnitude lower in the Central German Bight (0.04–1.37 ng L^{-1}). The geographical distribution of this compound seems to indicate a dilution from higher concentrations in the south to the northwestern sampling stations. This may be due to the fact that the river

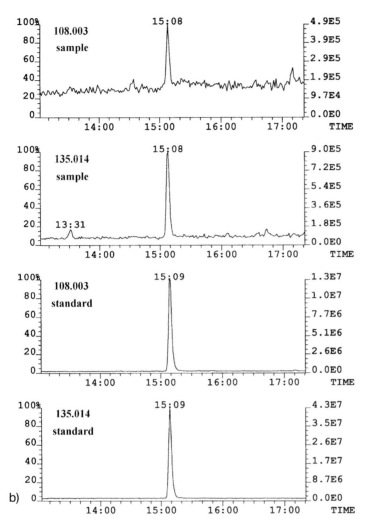

Fig. 2.88 (b) Ion fragment chromatograms (in SIM mode) of masses 108.003 and 135.014, significant for benzothiazole from station 4 taken between the island of Sylt and the central German Bight (upper two chromatograms) and a standard (lower two chromatograms). (Reprint with permission from [60]).

is an important source of immissions of these pollutants as well as other rivers that reach the German Bight at its southern coast.

Benzothiazole is more uniformly distributed in the waters of the German Bight than is MTB. Even in the estuary the concentrations did not exceed 2.7 ng L^{-1}, while they comprised 0.66 ng L^{-1} in the central German Bight in 1990. However, in 1995 some values were near or below the blanks determined in this study. Ob-

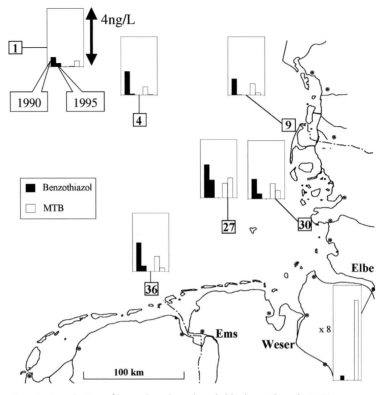

Fig. 2.89 Distribution of benzothiazole and methylthiobenzothiazole (MTB) in water of the German Bight of the North Sea. Data are given for 1990 and 1995. Maximum values are 4 ng L^{-1} within the fields. The rivers Ems, Weser, and Elbe are indicated. (The bars in the Elbe estuary are reduced to one-eighth scale to make them fit the graph.)
(Reprint with permission from [60]).

Table 2.46 Concentrations (ng L^{-1}) of benzothiazoles in the German Bight in 1990 and 1995.

	Benzothiazole 1990	Benzothiazole 1995	MTB 1990	MTB 1995
Sample station				
1	0.66	(0.24) [a]	0.04	0.39
4	1.61	(0.08) [a]	0.55	0.044
9	1.11	(0.03) [a]	0.78	0.16
27	2.27	1.23	1.00	1.37
30	1.38	0.37	1.06	0.60
36	1.97	0.40	1.04	0.24
Stade/Elbe	Not analyzed	2.74	Not analyzed	55.00

[a] Values are near or below the blank value.

viously, the Elbe River contributes little to the concentrations of benzothiazole in the German Bight. If another river such as the Rhine were the source of benzothiazole in the German Bight, the concentration pattern would be more conceivable. It is interesting to note, though, that at nearly all sampling sites the concentrations of both benzothiazole and MTB were lower in 1995 than in 1990. Whether this was a seasonal or hydrodynamic effect or was a result of diminished inputs can be answered only by more-detailed future studies.

2.6.1.3 Discussion and Conclusions

The concentrations found for both benzothiazole and MTB are of the same order of magnitude as those for α- and γ-HCH [61], thiophosphates such as O,O,O-trimethylthiophosphate and O,O,S-trimethyldithiophosphate [63], nitrobenzene [62], and polycyclic musk fragrances [39] in the German Bight, but they are considerably lower than those found for triazine herbicides in coastal waters (see Table 4.3) [64]. In contrast to the thiophosphates, which are mostly attributed to Elbe River inflow to the German Bight, the distribution of benzothiazoles seems to indicate inputs other than the Elbe, possibly other rivers (e.g., Rhine, Thames, Tee, Tyne, Ems, or Weser), or diffuse sources such as atmospheric deposition. This assumption is in accordance to the study of Welch and Watts [191], who found BT in atmospheric deposition samples. Because these compounds may be released from all kinds of rubber materials, e.g., automobile tires, diffuse sources should not be neglected.

The incomplete database concerning spatial and temporal resolution currently limits interpretation with respect to sources and temporal trends of the benzothiazoles determined in this study. However, the fact that these compounds could be observed at concentrations in the marine environment similar to other well-known contaminants justifies further studies.

Metcalfe et al. [190] reported K_{ow} values of 100 and 1000 for benzothiazole and MTB, respectively. The bioconcentration factors for leeches in freshwater obtained in their experiment ranged from 200 to 460 (benzothiazole) and from 100 to 470 (MTB). The half-lives of these compounds (1–2.5 days for MTB and 7 days for benzothiazole) in leeches were higher than the half-life of lindane (1 day). It is interesting to note that benzothiazole was also identified in fish samples [192]. Hendriks et al. [184] tested MTB and benzothiazole for acute toxicity to *Daphnia magna* and found EC_{50} values of 10 mg L^{-1} and 50 mg L^{-1}, respectively. Luminescence inhibition (EC_{50}) of *Vibrio fischeri* was measured by Reemtsma et al. [188] at 6 µg L^{-1} for TCMTBT and 3 mg L^{-1} for benzothiazole. The same group found an inhibition of nitrification in sediment columns at concentrations of 0.1–0.3 µg L^{-1} for benzothiazole and MBT as well as another derivative. Whether or not the concentrations of benzothiazole and MTB found in water from the North Sea could have any long-term effect on marine wildlife or the marine ecosystem is completely unknown at present. The presence of lipophilic and biocidic compounds such as benzothiazoles may be considered a cause for concern.

2.6.2
Enantioselective Degradation of Bromocyclene in Sewage Treatment Plants

2.6.2.1 Introduction

Bromocyclene (structural formula in Fig. 2.90) has been utilized as an insecticide against ectoparasites in sheep production as well as in pet care. In the urban areas of Germany such as the Ruhr megalopolis, applications in pet care dominated over that in sheep production. However, production of bromocyclene in Germany was stopped around 1995. Because of its high bioaccumulation potential, it was detected not only in sewage systems and sewage treatment plants [193] but also in freshwater fish [10, 194]. Enantioselective determination at that time was used to obtain results on the biodegradation of bromocyclene in fish. Considering the long time period since the phase-out of bromocyclene, it was surprising that it was easily identified in sludge samples from 2002 (see Figs. 2.91 and 2.92).

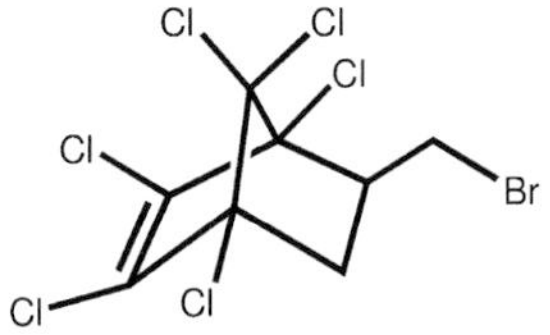

Fig. 2.90 Structural formula of bromocyclene.

2.6.2.2 Methods and Materials

Sewage sludge was sampled from 20 STPs in North Rhine-Westphalia (Germany). The respective sludge samples were lyophilized and homogenized, and subsamples of these were Soxhlet-extracted (6 h ethyl acetate). The extracts were concentrated afterwards and successively cleaned up by means of size exclusion and silica. (The procedure is described in detail in [38] as well as in Sections 2.1.1 and 3.3.) The samples were quantified by means of GC-MS in electron impact (70 eV) mode using a Trace GC-MS (Thermo-Finnigan, Dreieich, Germany) system equipped with a DB-5 column. The following temperature program was used: $90\,^{\circ}\mathrm{C}$ (2 min) $\rightarrow 10\,^{\circ}\mathrm{C}\ \mathrm{min}^{-1} \rightarrow 280\,^{\circ}\mathrm{C}$ (15 min).

The enantioselective determination was performed after an additional silica gel cleanup (eluent n-hexane 5% MTBE). The samples were analyzed by GC-MS utilizing a 25-m 0.25-mm i.d. heptakis-(2,3-di-O-methyl-6-O-t-butyldimethyl-silyl)-β-cyclodextrin column (Macherey und Nagel, Düren, Germany) and the following temperature program: $110\,^{\circ}\mathrm{C}$ (1 min) $\rightarrow 5\,^{\circ}\mathrm{C}\ \mathrm{min}^{-1} \rightarrow 152\,^{\circ}\mathrm{C}$ (40 min) $\rightarrow 5\,^{\circ}\mathrm{C}\ \mathrm{min}^{-1} \rightarrow 230\,^{\circ}\mathrm{C}$ (20 min) in a Trace Plus GC-MS system in electron impact (70 eV) mode (Thermo-Finnigan, Dreieich, Germany).

Baseline separation of the enantiomers was achieved and three mass fragment ions, i.e., 357, 359, and 361, were used for enantioselective determination (see Fig. 2.92).

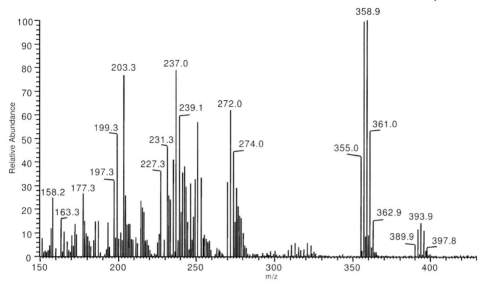

Fig. 2.91 EI mass spectrum of bromocyclene from a sludge sample extract.

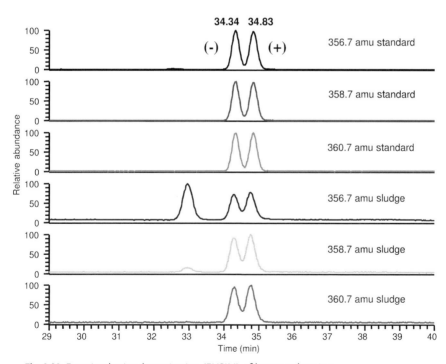

Fig. 2.92 Enantioselective determination (EI/SIM) of bromocyclene in a sewage sludge sample extract. Elution order following Bethan et al. [10].

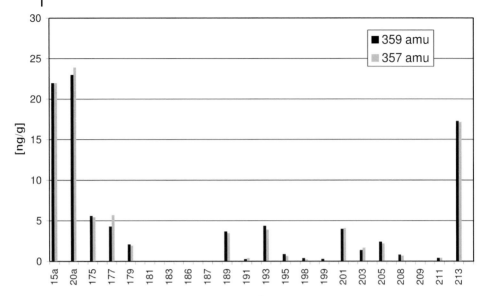

Fig. 2.93 Concentrations of bromocyclene (ng g^{-1} dry weight) in sewage sludge sample extracts (15 a and 20 a originate from the same treatment plant).

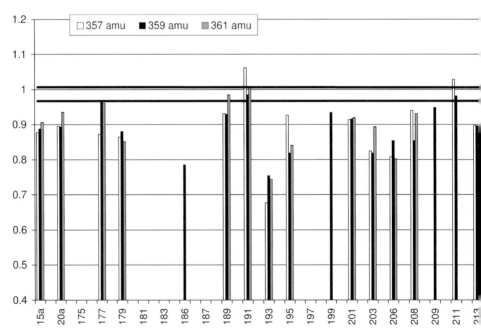

Fig. 2.94 Enantiomeric ratios of bromocyclene in sewage sludge sample extracts. The standard deviations of the measurements are indicated by a bar.

2.6.2.3 **Results and Discussion**

Bromocyclene was identified by GC-MS in German sewage sludge samples from 2002 (Fig. 2.91). In these samples the concentrations varied from <0.1 ng g^{-1} to 24 ng g^{-1} (Fig. 2.93). This variation is much higher than those determined for triclosan, the polycyclic musks HHCB and AHTN, and other compounds determined in the same samples. On the other hand, the concentrations of bromocyclene were significantly lower than those for triclosan. This leads to the assumption that either diverse processes are relevant in STPs or the sources are diverse and possibly not continuous in time.

Enantioselective analysis was applied to study whether or not biodegradation takes place in the respective sewage treatment plants, possibly in the digesters.

The enantiomeric ratios that were determined ranged from 0.75 to 1 (± 0.02) (Fig. 2.94). This may indicate that bromocyclene is not processed the same way in all STPs but that there are different ecosystems in the diverse digesters due to differing sludge retention, dominant carbon sources, temperatures, etc. These differences may result in different degradation pathways or kinetics.

However, the degradation process is not fast enough in any of the sampled STPs to eliminate bromocyclene totally from the sludge. Because 30% of sewage sludge is currently added to agricultural land, this may lead to increased concentrations of this biocide in soils. The emissions of bromocyclene from STPs have already led to elevated levels in fish.

3
Analytical Chemistry Methods

In this chapter the applied methods are described in more depth in general terms, and compound-specific information is given in the respective Results sections. Additionally, some method development results are described that exhibit value in respects other than the experiments described in earlier chapters.

3.1
Fresh and Wastewater

3.1.1
Lipophilic Compounds from Fresh and Wastewater (GC Analysis)

3.1.1.1 Sampling
Limnic and water samples were collected in 1-L glass bottles that were closed with screw caps with teflonized sealings.

Surface Water
The bottles were manually opened below the water surface and left there with the mouth open to the stream of water. When filled, they were closed below the surface before they were taken out of the water. This procedure was followed to make sure that a true representative of the water body was sampled and to exclude surface (oily) layers and sediment from the samples. These samples were stored at 4 °C until they were extracted, which typically took place the on the following day. The sampling strategy was to compare effluents with true river samples before and after discharges.

Wastewater
Wastewater samples were generally obtained as 24-h composite sample taken by means of automatic samplers with Teflonized connections. The samples were stored at 4 °C during the sampling interval. After 24 h a composite was produced, placed in 1-L glass bottles, and transported to the laboratory, where the extraction was performed the same day.

Personal Care Compounds in the Environment: Pathways, Fate, and Methods for Determination. Kai Bester
Copyright © 2007 WILEY-VCH Verlag GmbH & Co. KGaA, Weinheim
ISBN: 978-3-527-31567-3

3.1.1.2 **Extractions**

The extractions were performed as liquid–liquid extractions (LLE) with toluene in most cases, which proved to be a very robust method for wastewater analysis. Typical obstacles for wastewater analysis include high content of surfactants, fats, salts, and humic compounds and variations in pH. The respective recovery rates were not changed due to these parameters. This extraction proved to be superior to other methods in mechanical aspects (clogging, etc.) as well. In Fig. 3.1 it is demonstrated that pH has little effect on the recovery of even a phenolic compound such as triclosan when using this robust method.

In more detail, the unfiltered water samples were immediately extracted for 20 min with 10 mL toluene after adding an aliquot of internal standard solution (D_{15} musk xylene in toluene). The organic phase was separated from the aqueous one, and the water was removed by freezing the samples overnight at –20 °C. The crude extracts were concentrated on a rotary evaporator at 40 °C and 60 mbar to 1 mL. Final analysis was performed by GC-MS (Thermo-Finnigan Trace) equipped with a programmable temperature vaporizer (PTV) injector and a Trace autosampler. The PTV (1 μL injection volume) was operated in PTV splitless mode applying the following temperature program: 110 °C (0.1 s) $\rightarrow$ 10 °C s^{-1} $\rightarrow$ 310 °C (1.5 min) $\rightarrow$ 10 °C s^{-1} $\rightarrow$ 320 °C (10 min) (cleaning phase; open valve). The GC temperature program was 110 °C (1 min) $\rightarrow$ 40 °C min^{-1} $\rightarrow$ 165 °C $\rightarrow$ 4 °C min^{-1} $\rightarrow$ 210 °C 30 °C min^{-1} $\rightarrow$ 280 °C (15 min). The mass spectrometer was operated at 400 V on the photomultiplier and about 78 ms dwell time in selected ion (SIM) mode. The transfer line temperature was maintained at 230 °C, while the ion source was operated at 180 °C. The calibration was performed as a seven-step internal standard (D_{15} musk xylene) calibration (1–3000 ng mL^{-1} extract concentration) with 1/x weighting.

For verification purposes and for some analytes that were not extracted well from water by the LLE method, e.g., tris (2-chloroethyl) phosphate, a solid-phase extraction (SPE) method was applied. SPE is chronically vulnerable to clogging or even blocking as a result of high concentrations of particulate material as well as to

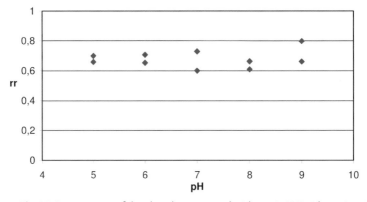

Fig. 3.1 Recovery rate of the phenolic compound triclosan in LLE with varying pH values.

changes in recovery rate due to high amounts of lipids, surfactants, and, in some cases, salt in the samples. To overcome blocking, etc., the geometry of the classical approach was changed. The classical approach is a cartridge with ~ 1 g sorbent in a 1-cm diameter device with 1 cm packing height (column length). In the experiments in this study, we used devices that had ~ 1 g packing material in a 5-cm diameter device (column diameter) but were only a few millimeters high. The packing was a highly efficient polymer packing (DVB-phobic) (Mallinckroth-Baker, Gross Gerau, Germany). These devices have been made commercially available after some work with homemade devices for marine samples (Section 3.2.2).

3.1.2
Steroid Hormones, Their Adducts, and Macrolide Antibiotics from Wastewater (HPLC-MS/MS Analysis)

3.1.2.1 Introduction

The occurrence of endocrine-disrupting chemicals as well as antibiotics in the environment has become an important issue in the last decades [195]. Steroid hormones and contraceptives are of special concern because of their endocrine potency [196]. The natural sex hormone estradiol, which has high endocrine potential, and its metabolites (estrone and estriol) and conjugates (glucuronides and sulfates) are excreted mainly in the urine of mammals [197]. The synthetic contraceptives ethinyl estradiol and mestranol also have high endocrine potential and are also excreted in the urine of women taking these drugs. Feminization of fish living near the effluents of sewage treatment plants has been observed [198, 199]; in addition, estrogenic effects on fish have been observed down to 1 ng L^{-1} in laboratory studies [200].

Antibiotics are also excreted by medicated humans, and they have the potential to create resistant strains of bacteria. In 1992, Neu [201] reported the resistance of bacteria to the majority of existing antibiotics. These compounds are transferred through the sewers to STPs and enter the environment through the plants' effluent.

Several methods have been described in the literature to quantify steroid hormones in STP effluents. They are based mostly on solid-phase extraction, derivatization of the analytes, and detection by GC-MS [202, 203]. The limits of quantification (LOQ) for such methods are in the low nanograms-per-liter range. The determination of the estrogens in unfiltered STP influents needs a sufficient cleanup step to remove interfering matrix components [204]. Hormone conjugates cannot be determined directly by GC-MS. These analytes must be enzymatically decomposed to gain the free estrogens prior to derivatization and successive analysis by GC-MS [205–207]. On the other hand, the conjugates can be determined directly by using liquid chromatography coupled with tandem mass spectrometry instead of GC-MS.

A method for the determination of roxithromycin, clarithromycin, and anhydro-erythromycin, a transformation product of erythromycin without antibiotic activity [197], in effluents of STPs was reported by Göbel et al. [208] and Hirsch et al. [209]. In these studies erythromycin was transformed to anhydro-erythromycin by using acidic conditions; thus, it remains unclear whether the samples contained the parent or the transformation product.

The aim of this study was to develop a reproducible, robust, and sensitive multi-residue method to investigate the fate of estrone (E1), 17β-estradiol (E2), estriol (E3), 16 α-hydroxyestrone (HE1), 17α-ethinyl estradiol (EE2), mestranol (ME), β-estradiol 17-acetate (E2Ac), β-estradiol 3-sulfate (E2S3), and estrone 3-sulfate (E1S3) roxithromycin (ROX) clarithromycin (CLA), and erythromycin (ERY) as its antibiotic-active form in influents and effluents of STPs by HPLC-MS/MS using the same sample and a single extraction and cleanup procedure. Such a multi-residue method is not described in the literature. The respective structural formulae are given in Table 2.39. Compliance within EU decision 657/2002/EC was desired for this project [210]. This recommendation essentially requests the analysis of two multiple reaction monitorings (MRM) per analyte to exclude false-positive results.

3.1.2.2 Experimental Methods

Materials
Water (HPLC grade) was obtained from Mallinckrodt Baker (Griesheim, Germany). Methanol, *tert*-butyl methyl ether (Suprasolv grade), ammonium acetate, acetone, ammonium hydroxide (analytical grade), and acetonitrile (Lichrosolv) were obtained from Merck (Darmstadt, Germany). Tetrahydrofuran (analytical grade) was obtained from KMF-Laborchemie (Lohmar, Germany).

Estrone, estrone 2,4,16,16-D$_4$, estriol, 16 α-hydroxyestrone, β-estradiol 3-sulfate sodium salt, and estrone 3-sulfate potassium salt erythromycin and roxithromycin were provided by Sigma-Aldrich (Seelze, Germany). 17 β-Estradiol hemihydrate, β-estradiol 17-acetate, 17α-ethinyl estradiol, and mestranol (Vetranal^{TM}) were obtained from Riedel-de Han (Seelze, Germany). Estrone 3-sulfate 2,4,16,16-D$_4$ sodium salt and 17 α-ethinyl estradiol 2,4,16,16-D$_4$ were obtained from Dr. Ehrenstorfer GmbH (Augsburg, Germany). Clarithromycin was provided by Promochem (Wesel, Germany). The synthesis of (E)-9-[O-(2-methyloxime)]-erythromycin is described in Ref. [211].

Internal Standards
Ten milligrams of 17α-ethinyl estradiol-d$_4$ (EE2-d$_4$) was dissolved in 90 mL acetone, and the volume was calibrated to 100 mL with HPLC water. Ten milligrams of estrone 3-sulfate 2,4,16,16-D$_4$ sodium salt (E1S3-D$_4$) was dissolved in 90 mL water, and the volume was calibrated to 100 mL with acetone. The internal standard solution for the steroid hormones (IS-H) was made by diluting these two stock solutions with methanol to a concentration of 10 ng μL^{-1}. The internal standard for the analysis of the macrolide antibiotics was made by dissolving 10 mg (E)-9-[-O-(2-methyloxime)]- erythromycin in 100 mL acetonitrile (IS-A).

Solid-phase Extraction

The DVB-phobic Speedisk cartridges from Baker (8086, Bristol, PA, USA) were conditioned with 15 mL methanol followed by 15 mL HPLC water. A solid-phase extraction manifold (IST, Grenzach-Wyhlen, Germany) with PTFE stopcock and outlet was used. The wastewater samples ($\sim$1000 mL) were passed through the cartridge at a speed of 100 mL min^{-1} (vacuum) by means of a Speedisk sample remote adapter (Baker, Bristol, PA, USA). The exact volume of the water sample was determined by difference weighting. The cartridge was washed with 15 mL HPLC water to remove ionic compounds and dried for 5 min by gently sucking air through the cartridge. The analytes were eluted from the cartridge with 15 mL *tert*-butyl methyl ether followed by 15 mL methanol in a 30-mL amber flask. Ten microliters of internal standard solution for the analysis of steroid hormones (IS-H) and 10 μL internal standard solution for the analysis of macro-lide antibiotics (IS-A) were added to the mixture. The flask was closed, shaken by hand for 10 s, and stored at –18 °C prior to cleanup.

Size-exclusion Chromatography Cleanup

Matrix components with high molecular masses were removed by size-exclusion chromatography (SEC). The wastewater extract was condensed to 1 mL at 60 °C and 35 hPa by a Büchi Syncore® Analyst 12-port evaporation unit (Essen, Germany). The residue was dissolved in 10 mL tetrahydrofuran (THF) and the volume was reduced again to 1 mL at 60 °C and 35 hPa before another milliliter of THF:acetone (70:30, v/v) was added. The complete sample was injected onto the SEC column. The SEC system consisted of a G1379A vacuum degasser, a G1311A quaternary pump equipped with a relay bus card (Agilent, Waldbronn, Germany), a rheodyne 7725i manual injection valve with a 2-mL sample loop (Rheodyne, Bensheim, Germany), and a C2-2006D automatic Valco valve (VICI AG, Schenkon, Switzerland) for fractionation. The size exclusion was performed on a Phenogel SEC column (i.d.: 21.2 mm, length: 300 mm, particle size 5 μm, 100) (Phenomenex, Torrance, CA, USA) at ambient temperature in an air-conditioned room at 24 °C.

Time synchronization between the injection and the fraction valve was triggered using a short contact closure signal of the injection valve to the pump. Also, the pump gave a short closure signal at time steps 21 min and 45 min to the fractionation valve.

A mixture of THF:acetone (70:30, v/v) was used as eluent at a flow rate of 3 mL min^{-1}. The first fraction from 0 min to 21 min, which contained the higher-molecular-weight compounds, was diverted to waste. The second fraction (21–45 min), which contained the hormones and antibiotics, was collected in a 100-mL amber flask. Afterwards, the column was rinsed for 5 min with the solvent prior to the next injection. The volume of the collected fraction was reduced to 1 mL by the Büchi evaporation unit at 60 °C and 35 hPa. The residue was dissolved in 10 mL methanol and the volume was reduced again to 1 mL at 60 °C and 35 hPa.

The best hard-cut and separation conditions were studied with MS/MS detection while the SEC-column was connected via a micro-splitter valve (P-460S,

Upchurch Scientific Inc., WA, USA) to the mass spectrometer (MS) and a split flow of 1 mL min^{-1} was introduced in the MS.

HPLC

The HPLC system consisted of a G1313A autosampler, a G1312A binary HPLC pump, a G1322A degasser, and a G1316A column oven (all from Agilent, Waldbronn, Germany).

Steroid Hormones and Conjugates

Separations were performed using a Synergi RP-MAX column (i.d.: 2 mm, length: 150 mm, particle size: 4 µm) and a SecurityGuard (Phenomenex, Torrance, CA, USA) at 25 ± 1 °C. The flow rate was 0.2 mL min^{-1}. The HPLC gradient was established by mixing two mobile phases (phase A: pure water; phase B: pure methanol). Chromatographic separation was achieved with the following gradient: 0–1 min: 0% B; 1–3 min: 0% to 70% B; 3–23 min: 70% to 100% B; 23–29 min: 100% B; 29–30 min: 100% to 0% B; 30–35 min: 0% B. Ten microliters of each sample were injected.

Macrolide Antibiotics

The HPLC separations of the macrolide antibiotics were performed using a Phenosphere-Next RP18 column (i.d.: 2 mm, length: 150 mm, particle size: 3 µm) and a SecurityGuard at 25 ± 1 °C. The flow rate was 0.2 mL min^{-1}. The HPLC gradient was established by mixing two mobile phases (phase A: 10 mM aqueous ammonia acetate solution; phase B: pure acetonitrile). Chromatographic separation was achieved with the following gradient: 0–1 min: 10% B; 1–14 min: 10% to 100% B; 14–29 min: 100% B; 29–30 min: 100% to 10% B; 30–35 min 10% B. Ten milliliters of each sample were injected. An additional methanol flow of 400 µL min^{-1} was added after the separation by means of a second G1312A binary HPLC pump to improve the ionization of the macrolide antibiotics in APCI mode.

Mass Spectrometry

The triple quadrupole mass spectrometer (API 2000, Applied Biosystems, Darmstadt, Germany) was equipped with a TurboIonSpray source (ESI) and a Heated Nebulizer source (APCI).

Steroid Hormones and Conjugates

APCI The APCI source operated under the following conditions: curtain gas (CUR): 35 psi; collision gas (CAD): 3 mTorr; nebulizer current (NC): 2 µA; temperature (TEM): 450 °C; ion source gas 1 (GS1): 60 psi; ion source gas 2 (GS2): 15 psi; interface heater (ihe): on; focusing potential (FP): 350 V; declustering potential (DP): 11 V; entrance potential (EP): 10 V.

The arrangement of the spray was vertical (v) 5 mm; horizontal (h) 3 mm to the orifice, and the position of the corona needle was v 4 mm and h 6 mm (Table 3.1).

Table 3.1 MS conditions and retention times for the analysis of steroid hormones with atmospheric pressure chemical ionization. Maximum variation in retention time was not larger than ±0.1 min (HPLC conditions: methanol: water, Synergi-RP Max).

	RT (min)	Precursor ion (amu)	Product ion (amu)	Dwell time (ms)	Collision energy (eV)	
HE1	12.4	287	251	100	21	v
		287	199	100	24	q
EE2	14.6	279	133	100	22	v
		279	159	100	27	q
EE2-D$_4$	14.6	283	135	100	28	q
		283	161	100	31	v
E1	14.9	271	159	100	29	q
		271	133	100	27	v
E1-D$_4$	14.9	275	161	100	29	
		275	135	100	32	
E2	14.8	255	159	100	24	v
		255	133	100	24	q
E2Ac	18.4	255	159	100	24	q
		255	133	100	24	v
ME	20.1	293	173	100	31	q
		293	147	100	27	v

q: MRM transition was used for quantification.
v: MRM transition was used for verification.

Electrospray Ionization
The ESI source operated under the following conditions: curtain gas (CUR), 40 psi; collision gas (CAD), 3 mTorr; ion spray voltage (IC), −4500 V; temperature (TEM), 400 °C; ion source gas 1 (GS1), 35 psi; ion source gas 2 (GS2), 70 psi; interface heater (ihe), on; focusing potential (FP), −350 V; entrance potential (EP), −10 V. The arrangement of the ESI spray to the orifice was v 5 mm and h 3 mm (Table 3.2). Both soft ionization modes were used in comparison.

Macrolide Antibiotics (APCI)
The APCI source operated under the following conditions: curtain gas (CUR), 50 psi; collision gas (CAD), 3 mTorr; nebulizer current (NC), 5 μA; temperature (TEM), 500 °C; ion source gas 1 (GS1), 80 psi; ion source gas 2 (GS2), 35 psi; interface heater (ihe), on; focusing potential (FP), 360 V; declustering potential (DP), 20 V; entrance potential (EP), 10 V.

The arrangement of the spray was v 5 mm and h 3 mm to the orifice, and the position of the corona needle was v 4 mm and h 6 mm.

MS calibration was performed up to m/z 1800 with mass resolution of quadrupole 1 and quadrupole 3; both were set to 0.7 Da. The data obtained were processed using Analyst^{TM} 1.4 software. To gain higher selectivity, MRM was cho-

Table 3.2 MS conditions and retention times for the analysis of steroid hormones with electrospray ionization in two time-controlled experiments: (1) t_1 = 0–13.5 min, (2) t_2 = 13.5–35 min. Maximum variation in retention time was not larger than ±0.1 min (HPLC conditions: methanol: water, Synergi-RP Max).

	RT (min)	Precursor ion (amu)	Product ion (amu)	Dwell time (ms)	Collision energy (eV)	Declustering potential (V)	
E1S3	11.6	349	269	100	−44	−68	v
		349	143	100	−100	−68	q
E1S3-D$_4$	11.6	353	273	100	−45	−68	q
		353	147	100	−76	−68	v
HE1	12.3	285	145	100	−50	−68	v
		285	159	100	−50	−68	q
E3	12.5	287	171	100	−49	−68	v
		287	145	100	−58	−68	q
EE2	14.6	295	145	100	−60	−80	v
		295	159	100	−47	−80	v
EE2-D$_4$	14.6	299	174	100	−60	−80	v
		299	161	100	−47	−80	v
E1	14.9	269	145	100	−47	−140	v
		269	159	100	−47	−140	v
E1-D$_4$	14.9	273	147	100	−51	−140	
		273	161	100	−51	−140	
E2	14.8	271	145	100	−54	−140	v
		271	239	100	−53	−140	v
E2Ac	18.4	313	253	100	−38	−68	v
		313	145	100	−59	−68	v

q: MRM transition was used for quantification.
v: MRM transition was used for verification.

sen. MS/MS parameters were optimized in continuous flow mode, injecting 1000 ng mL^{-1} standard solutions dissolved in methanol at a flow rate of 10 μL min^{-1}. The respective optimal collision energy (CE) and declustering potential (DP) were determined by means of a software procedure controlling the automatic switching between the different voltages with a step size of 1 V per scan and a range of −5 to −130 V in positive mode and 5 to 130 V in negative mode for the CE and 0–200 V in positive mode and 0 to −200 V in negative mode for the DP. The MRM transitions as well as the individual declustering potential and collision energy voltage used for the analysis of steroids hormones in ESI and APCI mode are displayed in Tables 3.1 and 3.2.

The macrolide antibiotics were analyzed more reliably in APCI, as demonstrated earlier for manure [211]. The parameters used in this study are demonstrated in Table 3.3.

Table 3.3 MS conditions and retention time for the analysis of macrolide antibiotics with atmospheric pressure chemical ionization. Maximum variation in retention time was not larger than ±0.1 min (HPLC conditions: acetonitrile: 10 mM aqueous ammonium acetate solution, Phenosphere-Next).

	RT (min)	Precursor ion (amu)	Product ion (amu)	Dwell time (ms)	Collision energy (eV)	
Clarithromycin	13.7	748	158	120	45	q
		748	590	80	25	v
Erythromycin	12.8	734	158	120	45	q
		734	576	80	25	v
Roxithromycin	14.2	837	158	120	46	q
		837	679	80	35	v
Internal standard (IS-A)	13.9	763	158	120	45	q
		763	605	80	25	v

q: MRM transition was used for quantification.
v: MRM transition was used for verification.

Calibration

The calibration was performed as an internal standard calibration. A stock solution for the hormones was produced by dissolving 10 mg of the hormones in 30 mL acetone and 30 mL water, and the container was filled up to 100 mL with acetonitrile. A stock solution for the macrolide antibiotics was produced by dissolving 10 mg of the respective antibiotics in 100 mL acetonitrile. These stock solutions were stored at 4 °C in the dark and were renewed after three months. Calibration standards (1, 5, 10, 50, 100, 500, and 1000 ng mL^{-1}) were made by serial dilution of the stock solution in methanol, and 10 µL mL^{-1} of each internal standard solution was added to the calibration standards. The calibration curves were calculated using a weighted (1/x) linear regression model.

3.1.2.3 Results and Discussion

All analytes were separated by HPLC. The calibration graphs were linear in the range from the limit of quantification (LOQ) up to 1000 ng mL^{-1} with correlation coefficients (r^2) better than 0.99.

Optimization of ESI Signals

The addition of buffers (ammonium acetate, ammonium formiate, or ammonium hydroxide at varying concentrations) to the mobile phase caused a decrease in the responses of the analytes due to lower ionization ratios for the separation of steroid hormones. The use of pure methanol instead of pure acetonitrile as phase B gave higher ionization ratios (by a factor of 2 to 3) in electrospray ionization mode for these analytes. These results correspond with the literature [207, 212]. The post-column addition of a 40 mmol L^{-1} methanolic ammonia solution gave a decrease in the response of the analytes in ESI negative mode. These results are different from those of Bartonti et al. [213] and Gentili et al. [214], who also used an API 2000 mass spectrometer.

Optimization of APCI Signals

The addition of ammonium acetate to the mobile phase of the macrolide antibiotics increases the ionization performance of these analytes. The addition of 400 μL min^{-1} methanol improved the ionization of the macrolide antibiotics. Thus, three HPLC-MS methods are necessary for an optimal ionization of all analytes.

Choosing an Internal Standard

The use of 2,4,16,16-D_4 estrone from Sigma-Aldrich as internal standard in an amount of 25 ng resulted in interferences with the corresponding undeuterated target compound up to concentrations of 3 ng L^{-1}, though Sigma-Aldrich claims an isotopic purity of 95%. For this reason, the use of 2,4,16,16-D_4 estrone was avoided in the final method.

The use of 2,4,16,16-D_4 estrone 3-sulfate sodium salt and 2,4,16,16-D_4 17α-ethinyl estradiol from Dr. Ehrenstorfer GmbH as internal standard gave no interference with the corresponding undeuterated target compounds. E1S3, E2S3, and E3 were quantified with sodium estrone 3-sulfate 2,4,16,16-D_4 as internal standard. 2,4,16,16-D_4 17α-Ethinyl estradiol was used to quantify E1, E2, EE2, ME, HE1, and E2Ac.

The use of (E)-9-[-O-(2-methyloxime)]-erythromycin as internal standard for the quantification of the macrolide antibiotics gave no blank problems.

Peak Identification

EU decision 2002/657/L221 requires four numbers of identification points for identification in LC-MS/MS analysis [210]. By using the HPLC-MS/MS technique, each precursor ion results in one identification point, while each transition product ion is counted as 1.5 points. Therefore, four points can be earned by measuring one precursor and two product ions. Also, the ratio of the chromatographic retention time of the analyte to that of the internal standard, the relative retention time of the analyte, should correspond to that of the calibration solution at a tolerance of ±2.5% [210].

Figure 3.2 shows the MRM chromatograms of ethinyl estradiol and its internal standard, D_4-ethinyl estradiol, in a standard solution and an STP influent sample. Using only the first MRM transition and the ±2.5% tolerance of the retention time criterion to identify the respective peak resulted in a false-positive identification. Only the use of a second MRM transition for verification gave a correct result.

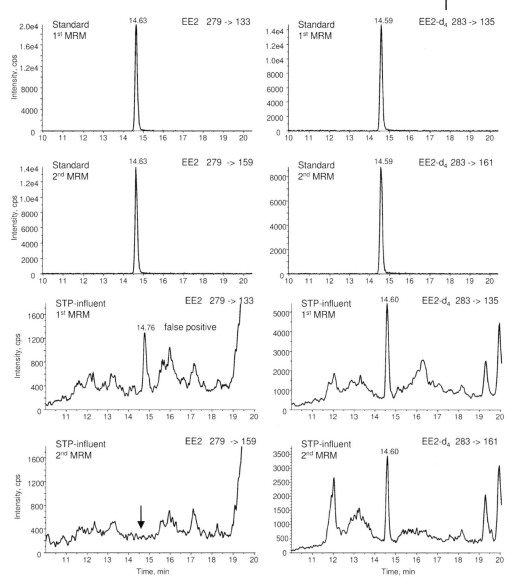

Fig. 3.2 MRM chromatograms of ethinyl estradiol and its internal standard, D$_4$-ethinyl estradiol, in a standard solution and an STP inflow sample. A false-positive determination is detected by the second MRM in the sample.

Matrix Effects

A major problem with quantitative liquid chromatography coupled with tandem mass spectrometry is the unexpected matrix effects. Different strategies have been established to compensate these matrix effects:

1. Matrix calibration: This is an internal standard calibration in the presence of an uncontaminated matrix to avoid matrix effects [216]. This matrix is produced by the same sample preparation that is used for the analysis of the samples. This matrix is successively added to the calibration standards. This method works well if there is access to an uncontaminated sample matrix. It is difficult to obtain matrix wastewater samples free of natural steroid hormones and antibiotics.

2. Standard addition: The sample is divided into several subsamples and a standard calibration solution is added to the subsamples [216, 217]. As a result, a calibration curve is generated for each sample. This method works well with a low number of samples, but it is impracticable with high sample throughput, as it causes multiplication of analysis time as the number of samples is multiplied by the number of standards.

3. Isotope dilution: This involves quantification with isotopic-labeled internal standards. These standards have the same chemical nature and co-elute with the respective analyte and the matrix. Thus, the same effect occurs to the internal standard and to the analyte. A disadvantage is the poor availability of these standards.

All three options compensate the matrix effects, but none of these options reduces these effects. Figure 3.3 shows the matrix effects of 2,4,16,16-D_4 17a-ethinyl estradiol at a concentration of 100 ng mL^{-1} with electrospray ionization (ESI). Despite an elaborate cleanup procedure, this internal standard shows 31 times less signal in an influent sample due to matrix effects, while a ninefold decrease is observed for effluent samples. As a consequence of these high matrix effects, quantification is difficult at concentration levels near the original LOQ. The limits of quantification are rising.

To reduce these matrix effects, chemical ionization at atmospheric pressure (APCI) can be utilized successfully. Figure 3.4 shows the chromatograms of 2,4,16,16-D_4 17a-ethinyl estradiol (EE2-D_4) in the same samples (standard, influent, and effluent) as discussed above for ESI. Compared to the standard, the ionization of EE2-D_4 in the influent and effluent are only a factor of 3.5 smaller, while in ESI this effect was a factor of 31. Thus, matrix effects in APCI mode are about a factor of 3–10 less pronounced than in the ESI mode. Similar effects have been observed for different matrices and ion sources from other manufactures in the literature [211, 216].

Thus, APCI is preferable to electrospray ionization for the analysis of steroid hormones in wastewater samples. Unfortunately, not all analytes could be ionized by APCI; therefore, E1S3, E2S3, and E3 were quantified by ESI-MS/MS. However, the matrix effects for these three hormones were much lower than for the later-eluting steroid hormones.

The comparison of ESI and APCI for the analysis of macrolide antibiotics as described in Ref. [211] revealed results similar to those obtained in this study focusing on hormones.

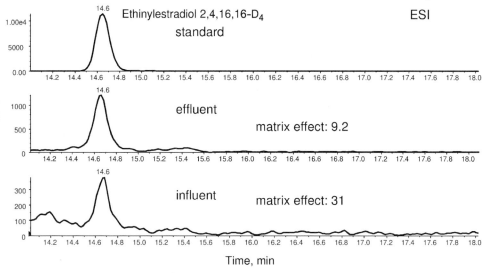

Fig. 3.3 ESI MRM chromatograms of 2,4,16,16-D$_4$ 17a-ethinyl estradiol in a standard solution in comparison to extracts of effluent and influent from STP samples (peak height) spiked to 100 ng mL^{-1} each. The resulting matrix effect is calculated by dividing the peak height (intensity) of the IS of the standard solution by the peak height of the IS of the respective sample extract.

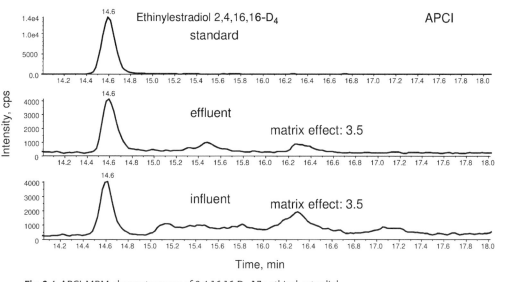

Fig. 3.4 APCI MRM chromatograms of 2,4,16,16-D$_4$ 17a-ethinyl estradiol in a standard solution in comparison to extracts of effluent and influent from STP samples (peak height) spiked to 100 ng mL^{-1} each. The resulting matrix effect is calculated by dividing the peak height (intensity) of the IS of the standard solution by the peak height of the IS of the respective sample extract.

Validating the Method

The method was validated primarily by spiking 1 L of tap water with the stock solutions to produce concentrations of 1, 3, 10, 30, 100, 300, and 1000 ng L^{-1} water. The following sample preparation, extraction, and cleanup were identical to the procedures described above. These recovery experiments for the hormones and antibiotics were carried out at seven concentration levels in triplicate.

The recoveries are given in Table 3.4. Because there was no significant concentration dependency of the recovery rate, the values of all experiments were averaged. The limit of quantification (LOQ) was defined as a signal-to-noise ratio of 10:1 and the limit of detection (LOD) as a signal-to-noise ratio of 3:1. The signal-to-noise ratios of the LOD and LOQ were taken from the chromatograms of the recovery rates.

Mean recovery rates of the steroid hormones in APCI mode ranged from 82% (RSD 12%) to 109% (RSD 15%). Mean recoveries of 58% (RSD 24%) for estriol and of 95% (RSD 13%) and 95% (RSD 12%) for the hormone sulfates were obtained using electrospray ionization. This method was also applied to β-estradiol 17-β-D-glucuronide, but it did not give constant recovery rates for this compound. The mean recovery rates for the macrolide antibiotics were 76% (RSD 14%) for roxithromycin, 82% (RSD 7%) for clarithromycin, and 100% (RSD 15%) for erythromycin (Table 3.4).

Stability of the Method with Respect to Matrix, Sample Transport, and Storage

To prove the stability of the analytical method, eleven 1-L STP influent samples were spiked with 100 ng of the respective analytes. Five spiked samples were extracted immediately, and five samples were stored in the dark at 4 °C for 48 hours before the extraction. The sample preparation, extraction, and cleanup of these 10 samples were identical to the procedures described above. The last sample was extracted directly after spiking, but this sample received no SEC cleanup. Another two unspiked influent samples were also analyzed to determine the background levels. Table 3.4 shows the recovery rates of the spiked influent samples, which were all identical to those obtained from spiked tap water within the precision of the method.

Sample Transport and Storage

In comparison with stored samples, degradation of only mestranol and 17 β-estradiol acetate was observable. The "recovery" was reduced from 97% to 68% for mestranol, while that of 17 β-estradiol acetate changed from 112% to 58%. Estradiol 3-sulfate shows a maximal "increase" of 10% of the concentration. The recovery rates of all other analytes were identical to those extracted immediately from wastewater and tap water.

Comparing Samples With and Without Cleanup

The different recovery rates of the crude sample (without SEC) in comparison to the cleaned samples were based on the different matrix effects of the internal

Table 3.4 Mean recovery rate, standard deviation, relative standard deviation (RSD), and limit of quantification (LOQ) (three extractions, repetitions for each concentration level) for hormones and antibiotics in tap water. Recoveries were determined at concentrations of 1, 3, 10, 30, 100, 300, and 1000 ng L^{-1} water. Also included are mean recovery rate ± standard deviation of five spiked STP influent samples, mean recovery ± standard deviation of five spiked STP influent samples stored for 48 h at 4 °C, recovery of one spiked influent sample without SEC cleanup, and limit of quantification (LOQ) of hormones and antibiotics in wastewater influents. LOQ: S/N = 10:1

	Mean recovery (%) in tap water $n=21$	RSD	Mean recovery (%) in influent $n=5$	Mean recovery after 48 h (%) in influent $n=5$	Recovery without SEC (%) $n=1$	LOQ (ng L^{-1}) in effluent and tap water	LOQ (ng L^{-1}) in influent
Hormones (APCI)							
Mestranol	105 ± 12	11%	97 ± 12	68 ± 7	73	3	6
16 α-Hydroxyestrone	74 ± 15	21%	86 ± 13	75 ± 8	87	8	8
17β-Estradiol	98 ± 9	9%	83 ± 8	88 ± 7	87	8	8
Estrone	105 ± 16	15%	90 ± 12	113 ± 11	64	2	4
17 α-Ethinyl estradiol	83 ± 6	7%	85 ± 9	82 ± 4	82	6	6
17 β- Estradiol acetate	109 ± 15	14%	112 ± 17	51 ± 6	104	1.5	3
Hormones (ESI)							
17 β-Estradiol 3-sulfate	92 ± 8	9%	95 ± 7	119 ± 11	98	1.8	28
Estrone 3-sulfate	95 ± 12	13%	77 ± 8	76 ± 7	74	0.6	4
Estriol	58 ± 14	24%	63 ± 19	43 ± 9	16	15	35
Antibiotics (APCI)							
Clarithromycin	82 ± 6	7%	91 ± 7	91 ± 12	30	2	2
Erythromycin	100 ± 15	15%	82 ± 9	81 ± 9	13	6	6
Roxithromycin	79 ± 8	11%	92 ± 9	86 ± 23	33	6	6

standards to their respective analytes. In comparison to samples that were processed with cleanup, those that were processed without cleanup revealed lower results for estrone, the macrolide antibiotics being analyzed by ESI using APCI, as well as estriol. The limit of quantification, defined as a signal-to-noise ratio of 10:1, in wastewater influent samples were maximally about two times higher than in tap water samples in APCI mode. In ESI mode, the LOQ increased by a factor of 7 (Table 3.4).

Application to Environmental Samples

The method was tested for several wastewater samples in order to investigate the fate of hormones and antibiotics during wastewater treatment. Wastewater from a sewage treatment plant in the Ruhr region of North Rhine-Westphalia (Germany) with an inhabitant equivalent value of 160 000 was sampled during four days. The samples were taken automatically as 24-hour composite samples at the inflow and effluent of the STP. The samples were refrigerated at 4 °C during the 24-h intervals. They were transported to the laboratory immediately after sampling and extracted within 6 hours after arrival. The samples were generally extracted on the same day. When it was not possible to extract the hormones immediately, the samples were stored at 4 °C for two days maximum. All samples were extracted in duplicates.

The steroid hormone 17β-estradiol was found at concentration of up to 22 ng L^{-1} in influents and of 8.6 ng L^{-1} in effluents (Table 3.5). The metabolites of 17 β-estradiol, estrone, 16 α-hydroxyestrone, and estriol were determined at maximum concentrations of 87 ng L^{-1} (E1), 90 ng L^{-1} (HE1), and 470 ng L^{-1} (E3) in influents and of 5.3 ng L^{-1} (E1), 14 ng L^{-1} (HE1), and 99 ng L^{-1} (E3) in effluents. These results were in the same level as those of Bartoni et al. [213] and Gentili et al. [214], who measured steroid hormones in the influents and effluents of different STPs in Italy. The conjugates of the steroid hormones were found in influents at maximal concentrations of 8–28 ng L^{-1} (E2S3) and 23 ng L^{-1} (E1S3). These results correspond with those of Gentili et al. [214], who found E1S3 at concentrations of up to 3.9 ng L^{-1}.

Estradiol 3-sulfate was found in effluents at maximal concentrations of 37 ng L^{-1} (E2S3), and estrone 3-sulfate was found at 14 ng L^{-1} (E1S3). These results match the results of Isobe et al. [218], who analyzed nine steroid hormone conjugates in the effluents of Japanese STPs and found only E1S3 and E2S3 at concentrations of 0.3–2 ng L^{-1}. The contraceptives mestranol, 17 α-ethinyl estradiol, and 17 β-estradiol acetate were detected neither in the inflow nor in the effluent samples of this STP. The concentrations of 17 α-ethinyl estradiol in wastewater were below the limit of detection (2 ng L^{-1}). Similar results concerning inflow and outflow data for estrone and estriol were obtained by Bartoni et al. [213]. However, Bartoni et al. [213] found 17 α-ethinyl estradiol at a concentration of up to 0.4–13 ng L^{-1} in the influents of Italian STPs.

Maximal concentrations of macrolide antibiotics in influents were found to be 370 ng L^{-1} (CLA), 160 ng L^{-1} (ROX), and 1200 ng L^{-1} (ERY). In effluents the concentrations of antibiotics were 230 ng L^{-1} (CLA), 130 ng L^{-1} (ROX), and 320 ng L^{-1} (ERY). Compared with the results another German group [209], who measured only anhydro-erythromycin, these values were in same range. In comparison with values from STPs in Switzerland [208], these concentrations were higher. Table 3.5 shows the concentrations in inflow and effluent samples of the hormones and antibiotics on four different sample days with different weather conditions.

3.1.2.4 Conclusions

A reliable multi-residue method with low LOQ has been developed to analyze steroid hormones and their conjugates, the synthetic contraceptives, and macrolide antibiotics unaltered in unfiltered influents and effluents of STPs. This method can be used to investigate the fate of these compounds in various steps of wastewater treatment.

Table 3.5 Concentrations of influents and effluents on four different sample days at an STP in the Ruhr region of North Rhine-Westphalia (Germany) with an inhabitant equivalent value of 160 000 and different weather conditions. The deviation is based on the relative standard deviation of the validated method.

		Day 1 (ng L^{-1})	Day 2 (ng L^{-1})	Day 3 (ng L^{-1})	Day 4 (ng L^{-1})
17β-Estradiol	Influent	18 ± 2	12 ± 1	11 ± 1	22 ± 2
	Effluent	–	–	8.6 ± 0.8	2.4–8 [a]
Estrone	Influent	45 ± 7	87 ± 13	42 ± 6	32 ± 5
	Effluent	4.6 ± 0.7	5.3 ± 0.8	–	2.0 ± 0.3
16a-Hydroxyestrone	Influent	13 ± 2	90 ± 14	18 ± 3	9.5 ± 1.4
	Effluent	2.4–8 [a]	6.7 ± 1.0	14 ± 2	6.8 ± 1.0
Estriol	Influent	54 ± 13	470 ± 110	66 ± 16	55 ± 13
	Effluent	20 ± 5	99 ± 24	4.5–15 [a]	–
β-Estradiol 3-sulfate	Influent	–	8–28 [a]	–	–
	Effluent	–	6.4 ± 0.6	3.0 ± 0.3	37 ± 3
Estrone 3-sulfate	Influent	1.2–4 [a]	23 ± 3	12 ± 2	5.3 ± 0.7
	Effluent	4.1 ± 0.5	3.0 ± 0.4	1.9 ± 0.3	14 ± 2
Clarithromycin	Influent	180 ± 12	370 ± 26	210 ± 15	81 ± 6
	Effluent	130 ± 9	230 ± 16	220 ± 16	81 ± 6
Erythromycin	Influent	180 ± 27	1200 ± 180	850 ± 130	97 ± 15
	Effluent	180 ± 27	320 ± 48	270 ± 41	66 ± 10
Roxithromycin	Influent	96 ± 13	160 ± 23	110 ± 15	54 ± 8
	Effluent	63 ± 9	110 ± 15	130 ± 18	38 ± 5

a) Value < LOQ.
–=< LOD; ethinyl estradiol < LOD (2 ng L^{-1}).

Solid-phase extraction followed by an SEC cleanup step resulted in sufficiently clean extracts, which were then analyzed by HPLC-APCI-MS/MS and HPLC-ESI-MS/MS. The electrospray ionization (ESI) was compared to atmospheric pressure chemical ionization (APCI) with regard to matrix effects in LC-MS/MS analysis. The APCI mode should be preferred over ESI, as it is less susceptible to matrix effects, even though less sensitivity is obtained in standard solutions.

For the analysis of hormones in wastewater, it is important to follow EU decision 2002/657/L221 to prevent false-positive results; thus, a second MRM should be used for verification.

3.2
Seawater

3.2.1
Lipophilic Compounds in Marine Water Samples

Analysis of seawater for organic micropollutants such as polycyclic musk compounds, chloroanilines, benzothiazole derivatives, etc., is quite difficult because the target concentrations are normally very low. Often, the concentrations are around 1 ng L^{-1} or even lower. Thus, reaching appropriate limits of quantification is quite challenging. One way to improve the signal-to-noise ratio in instrumental analysis is to increase the sample size. Therefore, 10- to 100-L samples are regularly processed for seawater analysis, while for limnic ecosystems a sample intake of 1 L is sufficient. On the other hand, blank problems concerning lipid films on the water surface that are known to accumulate lipophilic compounds are a threat to trace analysis. It is thus essential that the sampler pass the water surface in a closed condition and that it be opened at a determined depth. In these experiments this was performed by a 100-L round-bottom flask that could be opened at a determined depth. This setup also has the advantage that the sorption of lipophilic material to the glass surfaces is avoided, because the extraction takes place in the sampler itself. Thus, all surfaces are automatically rinsed with solvent during the extraction procedure. Special care has to be taken that extraction solvents are and stay clean. Additionally, the biological matrix in seawater may be the same as for limnic waters, especially during the times when marine algae blooms take place. Sample cleanup is necessary in these cases. In these experiments 100-L water samples were taken with an all-glass sampler according to Gaul and Ziebarth [219] and Theobald et al. [220]. Sampling was performed from aboard the research vessel *Gauss* at seven stations in the German Bight from a depth of 5 m in both 1990 and 1995. Prior to their extraction with 1 L *n*-pentane, immediately after sampling a solution of deuterated internal standards (3 *n*-alkanes, 10 aromatic hydrocarbons) was added to the water samples. One of them, D$_8$ naphthalene, was successively used as internal standard for the quantification. The extracts were dried with Na$_2$SO$_4$ and concentrated to 1 mL.

For non-target identification purposes, this extract was further condensed to 0.1 mL with a gentle nitrogen stream and fractionated with normal-phase high-performance liquid chromatography (NP-HPLC) afterwards. A ternary pump (Merck L6200, Darmstadt, Germany) was coupled to a Hewlett Packard 1050 autosampler and a Hewlett Packard 1040 diode array detector (Hewlett Packard, Waldbronn, Germany). A 25 cm×4 mm (5 µm) nucleosil column was first eluted with a gradient of *n*-hexane:dichloromethane 95:5 (2 min); then from 2 min to 10 min, it was programmed to 100% dichloromethane and held for 2 min. Additionally, the column was eluted with acetone for several minutes. This method has been described by Theobald et al. [153, 220] in detail. The samples were condensed to 0.1 mL after fractionation.

For quantification purposes the crude *n*-pentane extracts were concentrated with a rotary evaporator (1 mL) and cleaned up with 1-mL silica gel columns (SPE cartridge) with the eluent *n*-hexane:dichloromethane 10:1 for the first fraction and dichloromethane:*iso*propanol 1:1 for the second. No further fractionation was necessary. The respective eluates were condensed to 0.5 mL. Blanks were measured by extracting a sampler during the experiment with 1 L *n*-pentane. The respective extract was processed the same way as the water extracts used for quantification. This method proved to be very good for all lipophilic compounds such as organochlorines, polycyclic musks, thiophosphates, and nitroaromates. Triazine herbicides can be extracted the same way, if a suitable solvent such as ethyl or propyl acetate is chosen.

Quantification and identification were performed with a VG-Tribrid GC-MS system (VG-Micromass, Manchester, England). For identification purposes the gas chromatograph, a HP 5890 of the Tribrid system, equipped with an MN 52 (similar to DB 5) (length: 25 m; i.d.: 0.30 mm; film: 0.50 μm; carrier gas: He) obtained from Macherey und Nagel (Düren, Germany) was programmed from 40 °C 5 °C min^{-1} → 320 °C. A Gerstel cooled injection system was used with 4 μL injection volume. For identification purposes the mass spectrometer was operated at resolution m/Δm 500, the ion source was set at 180 °C, 70 eV, in electron impact mode, while scans from 50–400 *m/z* were performed with a cycle time of 1.2 s. During these measurements perfluorotributylamine (PFTBA) was used for the mass calibration. For quantification purposes the mass spectrometer was tuned to resolution m/Δm 1000 and selected ion mass fragmentograms were analyzed.

3.2.2
Hydrophilic Compounds in Marine Water Samples

A second method has been made available that is suitable for less lipophilic compounds. This method relies on high-volume solid-phase extraction of the respective seawater samples.

Solid-phase extraction (SPE) has gained major importance in the analysis of aqueous environmental samples. A broad variety of organic chemicals have been enriched from drinking [221, 222], ground [223, 224], waste [225, 226], surface [227–229], estuarine [230, 231], coastal [232], and marine waters [233, 234] by SPE. So far, emphasis has been placed upon organochlorine pesticides [235], organophosphorus and nitrogen pesticides [236, 237], chlorophenols [238], explosives [239], phthalates [240], aromatic sulfonates [241, 242], and PAHs [243]. The sorbents used in SPE include graphitized carbon black (GCB) [221, 244, 245], reversed-phase (RP) materials (modified silica gels), and polymeric materials. The most widely used RP material (and SPE sorbent in general) is the octadecyl (C$_{18}$) phase, but ethyl, butyl, cyclohexyl, octyl, phenyl, propylamino, dimethylaminopropyl, and cyanopropyl reversed phases have been applied as well [246–248]. With respect to polymeric sorbents, the best known are styrene–divinylbenzene copolymers (Polysorb S, Amberlite XAD-2, and XAD-4) and polyacrylates (Amberlite XAD-7 and XAD-8). Unsatisfactory recovery rates [246] and poor reproducibility [249] were observed for XAD resins. Especially for the XAD resins, excessive cleaning procedures are required prior to their use [250]. The development of a new generation of polystyrene-based sorbents overcame the problems associated with XAD resins. However, the capacity of these styrene–divinylbenzene and divinylbenzene–ethylvinylbenzene copolymers has risen immensely. It is specified to be 10-fold higher than that of C_{18}-RP sorbents. These polymeric sorbents have been used successfully for the extraction of the whole range of organic contaminants [242, 251, 252]. They proved to be especially suitable for medium to highly polar substances, where they showed substantially higher recovery rates than RP sorbents [225, 253] or liquid–liquid extraction (LLE). Even acidic pesticides such as dicamba, 2,4,5-trichlorophenoxy acetic acid (2,4,5 T), dinoterb, [254] and chlorophenols [255] were extracted quantitatively without acidification of the sample.

Similar to wastewater, clogging problems are important in marine analysis because high volumes have to be extracted to analyze low concentrations. In marine analytical chemistry, standard SPE methods can be applied only to estuarine or highly polluted coastal water samples with a volume of around 1 L. In the open sea, concentrations of most organic pollutants are low compared to limnic systems. Concentrations are typically in the lower nanograms-per-liter range (e.g., lindane [61]) or even in the low picograms-per-liter range (e.g., PCBs [256]). A conceivable possibility to meet the requirements for low detection limits is to raise the volume of the sample to 10 L, 100 L, or higher. Basic needs for large-volume SPE are (1) efficient online filtration, (2) high flow rates (to keep the extraction time within acceptable limits), (3) low flow resistance (of both the filter and the extraction unit), and (4) mechanical stability of the sorbent package.

Commercially available standard SPE systems are often incapable of handling these volumes. However, some approaches to solid-phase extraction of large-volume (>10 L) seawater samples have been reported in the literature. For example, Gómez-Belichn et al. [257] extracted hydrocarbons, PAHs, PCBs, and fatty acids from seawater with XAD-2, PUF, and by LLE. Volumes ranged from 42 L to 1000 L at flow rates of 100–1900 mL min^{-1}. Ehrhardt and Burns [258] extracted 221–435 L of coastal and inshore waters with a Seastar Infiltrex sampler and 100 mL XAD-2 resin for the analysis of *n*-alkanes, alkylbenzenes, and PAHs. Schulz-Bull et al. [256, 259] continuously filtered and extracted seawater for analysis of PCBs from aboard a ship at a flow rate of 500 mL min^{-1} using GF/C filters (120 mm) and XAD-2-filled glass columns (100 mL and 120 mL), where extracted volumes ranged from 50 L to 1140 L. The same group developed a submersible in situ filtration/extraction system (KISP) [260]. The system was tested at flow rates between 1250 mL min^{-1} and 1500 mL min^{-1} with a XAD-2 volume of 250 mL and extracted water volumes of 210–700 L. The limiting factor, apart from energy supply, was found to be the clogging of the filter with suspended particulate matter (SPM). Sturm et al. [261] extracted samples of

100 L of coastal water with an Infiltrex II sampler and 30 g of C_{18} material for the determination of PAHs and organochlorines. Generally, the loaded sorbents were back-extracted in a Soxhlet or Ehrhardt apparatus, which is time-consuming and prone to contamination and requires considerable amounts of solvents.

Most work in marine analytical chemistry focuses on classical target compounds such as organochlorines and PAHs. Large uncertainties remain about the inventory of other, especially polar, organic pollutants in marine systems. This results in the need for a method to extract and identify a broad variety of substances. Within this work a sampler was constructed that allows the online filtration and extraction of large volumes (10 L) of coastal and offshore waters at high flow rates. Furthermore, an analytical method for the application of this filtration/extraction unit to non-target screening of seawater for a wide range of organic pollutants was developed. In contrast to the cited large-volume methods, more modern principles of SPE were adopted (i.e., glass cartridges packed with relatively small amounts [2 g] of sorbent were used and the trapped analytes were directly eluted from the cartridge). Thus, the method extends the applicability of common SPE procedures to large-volume samples and, by the use of the polymeric sorbent, to highly hydrophilic compounds.

For an overall assessment of organic pollutants in aquatic ecosystems, it is crucial to set up inventories of the chemicals present in the respective compartment by a non-target screening. This has been done rather completely for the Elbe River and its tributaries [262]. Further investigations have been carried out, e.g., for effluents from Swedish STPs [27], for the Elbe estuary [153], and some toxicity identification evaluation (TIE) approaches (e.g., in the Rhine delta [184]).

However, no systematic experiment based on this approach has been reported for the North Sea, although results for single "non-target compounds" have been published [166]. In this work, the results of non-target screening of North Sea water performed with the newly developed method will be presented.

3.2.2.1 Experimental Methods

The filtration and extraction units were manufactured by the mechanical workshop of the Department of Chemistry, University of Hamburg, from stainless steel 4301 (Schoch, Germany) and PTFE (Wegener, Hamburg, Germany), respectively. The glass cartridges were made by the institute's glassblowers from glass tubes and glass frits (pore size 40–100 µm) (Winzer, Germany). Glass fiber filter candles (exclusion size 0.5 µm) were obtained from Voigt (Wernau, Germany). They were cleaned prior to filtration by heating for 72 h at 420 °C and duplicate Soxhlet extraction (200 mL *n*-hexane/ethyl acetate, 6 h). Glass fiber filter disks GF/C (diameter 47 mm, exclusion size 0.45 µm) were supplied by Whatman (Maidstone, UK). The commercially available sorbent Bakerbond SDB-1 (styrene–divinylbenzene copolymer, particle size 40–120 µm, pore size 27 nm) was obtained from Baker (Griesheim, Germany).

Chemicals for the recovery experiments and verification of the non-target screening results were obtained from Merck, Darmstadt, Germany (2,5-dichloroaniline, dichlorobenzenes, caffeine, chloroaniline, chloronitrobenzenes, di-*n*-butyl phthalate, *N,N*-diethyl-3-toluamid (DEET), nitrobenzene), Sigma-Aldrich, Steinheim, Germany (3-chloro-4-fluoronitrobenzene, 2,4-dibromoanisol, dichloro-pyridines, triphenylphosphinoxide), Promochem, Wesel, Germany (desethylatrazine, HCHs, hexachlorobenzene, octachlorostyrene, parathion-methyl, pirimicarb, propoxur, terbutylazine), Synopharm, Barsbüttel, Germany (carbamazepine, propyphenazone), Dr. Ehrenstorfer, Augsburg, Germany (atrazine, desethylterbutylazine, dimethoate, metolachlor, simazine, trichlorobenzene), Fluka, Neu Ulm, Germany (tri-*iso*-butyl phosphate, tri-*n*-butyl phosphate) and ABCR, Karlsruhe, Germany (2,6-dichlorobenzonitrile=dichlobenil). All solvents used were of organic trace analysis grade and were supplied by Merck, as was HPLC-grade water. Sodium sulfate (granulated, organic trace analysis grade, Merck) was heated for 6 h at 600 °C) before use. For fractionation, columns were prepared from glass cartridges (8 mL), PTFE frits (pore size 20 μm), and 2 g silica (particle size 40 μm, pore size 6 nm, analytical grade) (all from Baker, Griesheim, Germany). The silica was activated prior to use by heating for 12 h at 120 °C. Ultrapure water was prepared with a Seral-Pur Pro 90C apparatus (Seral, Ransbach, Germany).

The standard stock solution of ca. 200 mg L^{-1} was prepared by dissolving about 20 mg (range 12–64 mg) of the pure compounds in 100 mL acetone. The working standard solutions were obtained by further dilution with acetone (for spiking) or *iso*-octane (for external quantification). All solutions were stored at 4 °C in the dark.

Water samples were taken using a 10-L glass sphere sampler as described by Gaul and Ziebarth [219]. Two samples from the Elbe were taken for recovery and screening studies on 23 March 1998 at Hamburg-Neumühlen (km 627, right bank, depth: 2 m). The North Sea water sample used for screening purposes was taken from aboard the research vessel RV *Gauss* on 27 June 1998 at the position 54° 13.50' N, 8° 23.00' E, sampling depth 5 m below the water surface.

After sampling, the water was pumped from the sampler via PTFE tubing through the filtration/extraction unit by a gear pump (model MCP-Z, pump head Z-120 with PTFE gears, magnet 66; Ismatec, Wertheim, Germany). For the determination of recovery rates from purified water, the pump was placed behind the extraction unit. The pump had to be placed between the filtration unit and the extraction unit for the handling of environmental samples with a high load of particulate matter, because with rising flow resistance, air was drawn in at the exit of the extraction unit when a vacuum approach was used. This led to malfunctioning of the pump. Situating the pump before the extraction unit did not cause contamination as checked by procedural blanks.

For online filtration of the sample, glass fiber filter candles (height 82 mm, outer diameter 26 mm, inner diameter 14 mm) were used in a stainless steel housing (see Fig. 3.5).

Extraction was performed in a specially designed and constructed device (see Fig. 3.6). A glass cartridge (height 57 mm, inner diameter 45 mm) was inserted

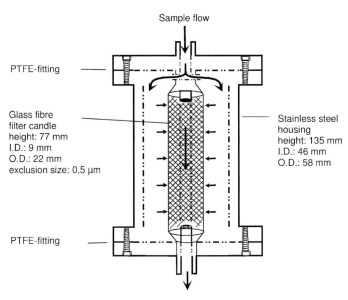

Fig. 3.5 Filtration unit for use with filter candles.

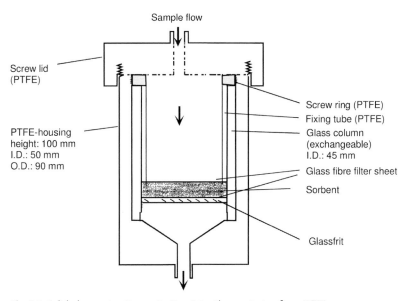

Fig. 3.6 Solid-phase extraction unit. (Reprint with permission from [272]).

into the PTFE housing. The bottom was covered with a glass fiber filter sheet (GF/C). Two grams (ca. 6 mL) of the sorbent were filled in a suspension in *n*-hexane and covered with another filter sheet, resulting in a bed height of approximately 5 mm. The package was fixed for mechanical consistency by introducing the PTFE cylinder and was locked by the screw ring. Uniform thickness of the sorbent layer was controlled by a fixing screw (see Fig. 3.6). The sorbent was rinsed with 50 mL *n*-hexane and 50 mL ethyl acetate by application of aspirator vacuum at the exit of the extraction unit. The column was conditioned by passing through 50 mL of methanol and 50 mL of HPLC-grade water.

After having filled all parts of the filtration/extraction unit with water, it was connected to the pump and the sample was pumped through the experimental set at a flow rate of 500 mL min^{-1}. After the extraction step, the wet extraction cartridges and filter candles were removed and stored in screw-top glasses at $-18\,^{\circ}$C until elution.

After defrosting, the loaded cartridges were inserted and fixed in the extraction unit (Fig. 3.6) as described above. The solid phase was eluted by three portions of 30 mL ethyl acetate, followed by 50 mL *n*-hexane/ethyl acetate (4:1). A gentle vacuum was applied to suck the solvent through the cartridge, and the combined eluates were collected in a 250-mL round-bottom flask. Relatively high amounts of ethyl acetate were used for elution in order to repel residual water from the sorbent. This approach avoided excessive drying of the cartridge and thus reduced potential contamination or losses of analytes. After phase separation the aqueous phase (ca. 5 mL) was pipetted off and re-extracted twice with 1 mL *n*-hexane. The combined organic phases were dried over sodium sulfate. The solvent was pipetted off and the residual sodium sulfate slurry re-extracted three times with 10 mL *n*-hexane/ethyl acetate (4:1). The combined organic phases were reduced to a final volume of approximately 150 µL in a rotary evaporator after addition of *iso*-octane as a keeper.

The filter candles loaded with SPM were not extracted because the main interest of this work was focused on hydrophilic analytes. However, extraction in a Soxhlet apparatus can be performed, thus including the organic fraction adhering to particles.

No cleanup was necessary for the recovery studies from purified water, and thus it was not performed in this case. The intention was to check the performance of the new SPE approach rather than that of the cleanup. This means that the recovery rates presented herein do not include any cleanup steps.

In contrast, a cleanup improved the proper identification of trace organics in the course of non-target screening of environmental samples. In this case, a silica cleanup/fractionation similar to that described by Specht and Tilkes [263] was applied. In an 8-mL glass cartridge, 2 g silica were packed between two PTFE frits and conditioned with 10 mL *n*-hexane. After application, the sample was eluted with 6-mL portions of first *n*-hexane and then *n*-hexane:dichloromethane 9:1 (v/v), *n*-hexane:dichloromethane 4:6 (v/v), dichloromethane, dichloromethane:ethyl acetate 1:1 (v/v), ethyl acetate, acetone, and finally 12 mL methanol. The eluates were reduced to a final volume of approximately 150 µL in a rotary evaporator after addition of *iso*-octane as a keeper.

For the recovery studies and non-target screening of environmental samples, a Magnum ITD (Finnigan MAT, Bremen, Germany) ion trap mass spectrometer was used under the following conditions: EI ionization at 70 eV, manifold temperature 200 °C, emission current 10 µA, scan range 100–420 amu. It was coupled to a Varian 3400 GC (Varian Associates, Sunnyvale, USA) (split/splitless injector 1075, 60 s splitless, 250 °C; column DB5-MS; J&W Scientific, Folsom, USA), length 25 m; i.d.: 0.20 mm; film thickness: 0.33 µm; carrier gas helium 5.0 [75 kPa]; transfer line 250 °C) run with an A 200 SE autosampler (CTC Analytics AG, Zwingen, Switzerland) (injected volume 2 µL). Temperature programs were usually (60 °C) (2 min) $\rightarrow$ (10 °C min^{-1}) $\rightarrow$ 260 °C (20 min) for recovery studies and 60 °C (2 min) $\rightarrow$ (7 °C min^{-1}) $\rightarrow$ 260 °C (20 min) for environmental samples.

3.2.2.2 Results and Discussion

Mechanical Performance of the Filtration/Extraction Unit

The use of filter candles in the filtration unit proved to be highly effective. Because of the high capacity of the filter candle as compared to the usual membrane filters, clogging was not relevant. Even samples of 10 L from the Elbe, containing high loads of suspended particulate matter (typically 20–40 mg L^{-1}), were filtered with the system at a routine flow rate of 500 mL min^{-1}.

The construction of the extraction unit combines the advantages of extraction disks and extraction columns. The high, disk-like diameter lowers the flow resistance. The column-like packing of the sorbent not only raises the overall capacity of the cartridge but also decreases the risk of breakthrough. The extent of breakthrough for a given sorbent is directly dependent on the polarity of the analytes, the extracted volume of water, and the sorbent bed height. Furthermore, a sufficient bed height (capacity) is especially important for surface and seawater, because dissolved organic carbon competes with the analytes for adsorption. Compared to common SPE approaches, the achieved flow rates are exceptionally high without decreasing recovery. This is a product of the efficient online filtration achieved by filter candles in contrast to membrane filters, the diameter of the extraction cartridge, and the hydrodynamic and sorptive properties of the solid phase used. C$_{18}$ material would have required the use of 20 g of sorbent to reach the same capacity as 2 g SDB-1 and would have exhibited a higher flow resistance.

During the processing of river water samples with a high SPM load, an increase in the flow resistance was observed, probably caused by the advancing clogging of the filter candle and/or the filter sheet covering the sorbent. This resulted in the entering of air at the connection between the extraction unit and the pump. For this reason, the pump was placed between the filtration unit and the extraction unit for the processing of environmental samples.

Solid-phase Extraction

Recovery experiments were executed to determine the performance, reproducibility, recovery rates, and linearity of the method (Table 3.6). Because it was the aim of this work to develop a non-target screening method for seawater, a stan-

Table 3.6 Recovery rates and standard deviations (SD) of the method as determined from triplicate extractions from spiked purified water (10 L) at the 20 ng L^{-1} level.

Analyte	log K_{ow}	Recovery (%)	SD (%)
Octachlorostyrene	7.7	62	1
Hexachlorobenzene	6.4	67	4
Di-n-butyl phthalate	4.9	132	7
Tri-n-butyl phosphate	3.7	99	12
Tri-iso-butyl phosphate	3.5	80	8
Metolachlor	3.4	95	1
Terbutylazine	3.0	96	3
Parathion-methyl	3.0	95	2
2,5-dichloroaniline	2.9	86	17
Chlorobenzene	2.8	29	15
Nitrobenzene	1.8	66	6
Pirimicarb	1.8	98	3
Propoxur	1.6	107	10
Desethylatrazine	1.5	102	11

dard solution was used that represented a broad range of compounds. It included pesticides of different classes as well as industrial chemicals and additives (plasticizers), varying in polarity from octachlorostyrene (log K_{OW} 7.7) to dimethoate (log K_{OW} 0.7) [264, 265]. The compounds were chosen in such a way that co-elutions were avoided.

Ten-liter samples of purified water were spiked with 1 mL of the respective standard solution in acetone. They were pumped through the filtration/extraction unit at a standard flow rate of 500 mL min^{-1}, i.e., the extraction was completed within 20 min. Further treatment was carried out as described above. Quantification was routinely performed by GC/MS, except for dimethoate, which showed unsatisfactory chromatographic behavior and poor response on the GC/MS system. In this case, quantification was additionally performed by GC/PND.

Recovery rates and reproducibility were determined at an environmentally relevant concentration level ($\sim$20 ng L^{-1}) by triplicate extractions. The results are given in Table 3.6. Quantitative recovery was observed for polar analytes, whereas the highly nonpolar organochlorines were recovered in the 60–70% range. Interestingly, a good correlation was observed between polarity (given as log K_{OW}) and recovery of the analytes. This effect has to be attributed to either an irreversible sorption to the solid phase or incomplete extraction. Low recovery of chlorobenzene (29%) and, to a lesser extent, of nitrobenzene (66%) was due to their high volatility. Especially for chlorobenzene, evaporation procedures and solvent change to iso-octane led to severe losses and high standard deviation, which makes the method unsuitable for this compound. In the case of interest in highly volatile analytes, a respective modification of the elution and further treatment (e.g., elution with small volumes of dichloromethane) would raise recoveries. For dibutyl phthalate, a reliable determination was not possible because of contamination from the laboratory environment.

The recovery rates were checked for the influence of an environmental matrix by spiking and extracting a water sample from the Elbe (7.5 L, spiking level 800 ng L^{-1}). They were within the standard deviation of those from purified water for most analytes. Only octachlorostyrene showed a significant decrease in recovery (from 62% to 40%), which can be explained by its partitioning to suspended particulate matter (SPM) (whereas no decrease was observed for HCB). This behavior is well known for other highly chlorinated compounds, e.g., PCBs [256]. Thus, the missing amount might be found by extraction of the SPM-loaded filter candles. The investigation of the SPM load will be the subject of further works.

Linearity of the whole procedure was investigated at four points over a concentration range from 2 ng L^{-1} to 200 ng L^{-1} water (2, 10, 20, and 200 ng L^{-1}). The correlation between concentration and recovery was generally good (correlation coefficients 0.9994 or higher) for this range. Limits of quantification (LOQs) were estimated from the smallest peak area that could be quantified reliably, which corresponds to a signal-to-noise ratio of 10. The LOQs thus obtained for the instrumental performance were connected to recovery rates and sample volume to give overall LOQs for the entire procedure. They were in the range of 0.1 ng L^{-1} to 0.7 ng L^{-1} except for dimethoate (ca. 5 ng L^{-1}). It has to be pointed out here that this method was developed for non-target screening purposes in the full-scan mode of the MS. For quantitative investigations the given parameters (LOQ, RSD, linearity) would have to be determined more thoroughly for the analytes of interest. Additionally, other detection modes (e.g., SIM) would basically lower the limits of determination.

Screening of North Sea Water for Organic Compounds

After the present method had proven its aptitude for the investigated test compounds, its performance in non-target screening was tested with a sample of 8 L river water. The detection and identification of a large number of compounds, ranging from highly lipophilic organochlorines to readily water-soluble compounds such as caffeine, demonstrated its suitability for this purpose.

The newly constructed gear, as well as the co-developed analytical method, was then routinely applied to the extraction and screening of water samples from the German Bight and the entire North Sea. In order to demonstrate the power of this approach, the results from station DB30 will exemplarily be presented here (see Table 4.3). However, it should be noted that, facing the enormous number of micro-organics present in the North Sea, the intention was not to elucidate every peak/substance but to provide a fast screening for anthropogenic and potentially harmful substances. Identification was achieved by comparison of the obtained spectra with the NIST library. The large number of identified substances included PAHs, oxo-PAHs, alkylbenzenes, ethers, alcohols, aldehydes, ketones, esters, anilines, amides, nitro-compounds, N-heterocycles, sulfonamides, thiophenes, benzothiazoles, alkyl- and chloroalkyl phosphates, and various halogenated compounds such as chloroanilines and chlorinated bis-propylethers. Because identification by (low-resolution) spectra alone may lead to false assignments, library proposals for a couple of substances were verified by injection and measurement of the respective reference compounds. For some of the substances hitherto not reported to occur in the North Sea, concentrations were estimated by comparison with external standards of these compounds (not corrected for recovery rates). None of them were detected in the procedural blanks. Some of the verified compounds are listed in Table 4.3 and others are discussed in Sections 2.1.4, 2.4.6 and 2.6. Further research on the occurrence and distribution of the newly identified compounds in the North Sea is currently being carried out, as their presence, especially of those compounds known for their biological activity (pesticides, pharmaceuticals), poses a potential risk to marine ecosystems.

3.2.2.3 Conclusions

The SPE apparatus and the method reported herein were shown to be a valuable tool for the extraction of trace organics from large volumes of water, as typically required for the analysis of marine samples. Hitherto, the system had been tested for 10-L samples. An extension to higher volumes (e.g., 100 L) as well as higher flow rates seems possible and would further lower the achievable detection limits. The high diameter of the extraction cartridges and the small amount of sorbent (resulting in a relatively thin bed) in combination with the high-capacity filtration system allow high flow rates, which significantly lower the extraction time. In contrast to former approaches to large-volume solid-phase extraction in marine analytical chemistry, the analytes can be eluted directly from the cartridge. The polymeric sorbent used in this work proved to be apt for extracting analytes of a wide polarity range (log K_{OW} 7.7–0.7). It is especially suitable for medium to highly polar compounds (log K_{OW} range 3.5–0.7), which often are poorly recovered by LLE or C_{18} SPE. The possibility of extracting large volumes within a reasonable time enables high enrichment factors and thus very low limits of detection.

Application of the method to riverine and marine water samples demonstrated its strength in the analysis of organic compounds in environmental samples, even at concentrations in the picogram-per-liter range. Non-target screening of sample extracts from the German Bight revealed the presence of a large variety of potentially harmful substances, some of which are known as pollutants in river systems but have not yet been demonstrated to occur in the North Sea.

3.3
Sewage Sludges

Sewage sludges have generally been analyzed as final dewatered sludges after anaerobic stabilization. This material was part of the focus of this study, as it is exported from sewage treatment plants to be incinerated, to be deposited at dangerous-waste disposal facilities, or to be amended to soils (agricultural soils or for use in landscape architecture). It is a solid material containing 50–70% water and high amounts of organic carbon.

Because it was planned to establish a multi-method to determine nitroaromatic and polycyclic musks together with OTNE, organophosphorus flame-retardants, triclosan, potential metabolites, and bromocyclene and similar compounds, it was necessary to use a vigorous extraction (not a selective one) in combination with a powerful cleanup technology.

Real sludge samples from four different origins were analyzed in comparison with four extraction procedures. The sludges were two pressed (dry) sludges, liquid sludge to feed the digester (primary sludge), and sanitized sludge (calcium carbonate). The following alternative extraction methods were studied:

1. Soxhlet extraction with ethyl acetate of lyophilized sludge.
2. Soxhlet extraction with ethyl acetate of sludge dried with sodium sulfate.
3. Liquid–liquid extraction, first of the wet sludge (extracted with toluene) in a separation funnel, second of the sludge dried with sodium sulfate.
4. Lyophilized sludge samples extracted with ethyl acetate in comparison to humid sludge (60% water), which was extracted with acetone (both Soxhlet).

Extraction method 1 including lyophilization and successive Soxhlet extraction with ethyl acetate gave the highest concentrations and was thus used in all other experiments for sludge.

The cleanup was needed to tackle the high load of humic compounds as well as waxes, fats, etc. On the other hand, it could not be destructive, as several compounds would not withstand an oxidative or acidic cleanup. Thus, it was decided to use size-exclusion chromatography (SEC), also known as gel permeation chromatography (GPC), for cleanup purposes. SEC allows separation of small molecules (such as analytes) from larger ones such as humic compounds or fats. Size exclusion has been used extensively in food analysis as well as for sediments [263, 266]. The analysis of nitroaromates required very low detection limits (~ 10 ng g^{-1}), while the analysis of organophosphates was complicated by the high demands on cleanliness of injectors and pre-column. Thus, another cleanup step, i.e., an SPE on silica with ethyl acetate as eluent, was added.

The final method assessment was complicated because no uncontaminated sludges were accessible to spike them to known concentrations for recovery experiments. Thus, an artificial sludge was created from manure from a pig farm by mixing it with soil. Sewage treatment plants receive high loads of organic carbon such as feces and laundry wastes, but they also receive soil particles

Table 3.7 Recovery rates, standard deviations, relative standard deviations, and working range for several musk compounds and TCPP and triclosan from sludge.

Compound	rr (%)	SD (%)	RSD	Working range (ng g^{-1})
TCPP	115	41	0.36	100–1070
Triclosan	94	25	0.27	4–1270
Musk ambrette	106	22	0.21	3–10
HHCB	100	28	0.28	5–778
AHTN	76	13	0.17	6–1290
Musk xylene	81	13	0.16	2–1080
Musk moskene	97	16	0.16	1–10
Musk ketone	94	29	0.31	1–10

from street runoff as well as from ground water infiltration of the sewer system. Thus, the soil/manure mixture was believed to be as similar as possible to sewage sludge.

The sludge samples that were analyzed were obtained from the loading of the trucks. This was the final product after filtering, still containing about 60% water. These samples were stored at 4 °C until further processing in the same week of the sampling. They were lyophilized and 10 g of each sludge sample was extracted for 6 h with ethyl acetate in a Soxhlet apparatus. At the start of this process, an aliquot of the internal standard solution (D$_{15}$ musk xylene) was added. These extracts were transferred to toluene (rotary evaporator) in such a way that the samples were always covered with at least 1 mL solvent. The resulting extracts were cleaned up with 1 g silica SPE cartridges (silica 60 obtained from Merck) by elution with ethyl acetate. The resulting solutions were concentrated to 0.5 mL and injected to the GPC column (Biorad SX-3) 2.5×30 cm, flow 2.5 mL min^{-1} cyclohexane:ethyl acetate 1:1. The solvent eluting in the first 21:30 min was drained to waste, while the fraction 21:30–32 min was collected. Thus, macromolecules were separated as they eluted in the first fraction, while sulfur, etc., was separated from the target compounds as they were eluted after the analyte fraction as shown by Bester and Hühnerfuss [29, 266]. The samples were finally transferred to 1 mL toluene by means of a rotary evaporator. For the sludge sample extracts, the same GC-MS conditions were used as for the aqueous samples. Recovery rates were determined by spiking a 1:1 mixture of dried manure and soil with respective standard concentrations (Table 3.7).

The final concentrations were recalculated for moisture content of the sludge, which was determined in parallel by heating 1-g subsamples to 105 °C until constant weight was achieved (MA50, Sartorius, Göttingen, Germany).

4
Discussion

4.1
Sewage Treatment Plants

The basic setup of many current STPs is depicted in Fig. 4.1. In the aeration basin the organic carbon is transformed into sludge and carbon dioxide. Nitrogen (nitrate and ammonium) is transformed into elemental nitrogen. In the digesters the sludge is transformed anaerobically to some extent into methane. Some microcontaminants also are transformed. Halogenated compounds can be dehalogenated, and phenolic compounds can be methylated [267].

Currently, the quantitative effects of STPs on organic microcontaminants (with concentrations of less than 10 µg L^{-1}) are not fully understood. This study was performed to obtain more insight into this situation.

It was known that sorption to the sludge is an important process, although in the past "elimination" from the water phase was used as synonym for "degradation." In this study it was shown that sorption does occur, e.g., to the majority of the polycyclic fragrances and to some extent to triclosan. The most critical step for this is sludge–water separation in the settlement basin and in the final filter press/centrifuge. At the moment it must be assumed that the early stages in STP processes, such as the primary settlement basin, are most effective in eliminating organic compounds (see Section 2.1.1).

It was also known that in some cases transformation and possibly mineralization are indeed important processes. In this study it was shown that transformation is important for some part of the fate of triclosan and for the nitroaromatic musks and that some minor fractions of HHCB are transformed to the lactone. Oxidation processes, e.g., of HHCB to HHCB-lactone, take place mostly in the aeration basin. In the digester or the denitrification part of the activated sludge treatment, reductive processes dominate. Nitroaromatic musk fragrances, for instance, are reduced to the respective amines in the digester. Successively, the anilines are transported with the residual water stream into the aeration basin, from which they are emitted into the surface water because their sorption to sludge is rather poor.

On the other hand, this study has also shown that compounds such as chlorinated organophosphates, e.g., the flame-retardant TCPP, were not eliminated at all in the STPs that were studied.

Personal Care Compounds in the Environment: Pathways, Fate, and Methods for Determination. Kai Bester
Copyright © 2007 WILEY-VCH Verlag GmbH & Co. KGaA, Weinheim
ISBN: 978-3-527-31567-3

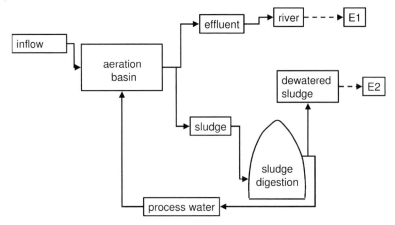

Fig. 4.1 Scheme of a typical STP. E1=emissions to the surface water; E2=emissions of sludge (generally to incineration plants, dangerous-waste landfill sites, or agricultural land).

For future optimization of STP performance, it will be helpful to use the knowledge generated in this study. However, a better understanding of the basic processes of elimination of organic micropollutants (xenobiotics) is still necessary. An overview of xenobiotics such as personal care products in the STP process is given in Table 4.1.

The following are management options for different elimination mechanisms:

1. Sorption mechanism: The compounds are mostly sorbed to sludge (biomass); primary sludge normally should contain higher concentrations than excess sludge. Management option: If sorption is the effective pathway of elimination, it will be the effective way to improve the STP's performance to increase the amount of sludge removed from the STP per day. This would end up reducing the sludge retention time. It will also result in more sludge to be treated in digesters. Decreasing sludge removal, e.g., in membrane bioreactors, will be counterproductive as they generate less sludge (up to no sludge) to be removed from the STP.

2. Oxidative biotransformation, especially in the aerated parts of the activated sludge treatment. Ideally, the final products of this process should be carbon dioxide, etc. Management option: Because this is normally a difficult process to perform on the more persistent xenobiotics, it might be interesting to increase the time in the aeration pathway, i.e., increase the hydraulic retention. However, these processes will often be triggered by more specialized bacteria; thus, higher sludge retention also would be efficient to perform this task. This is why membrane bioreactors sometimes are more efficient for chemically transforming the compounds in question (i.e., highly hydrophilic ones such as some pharmaceuticals).

Table 4.1 Overview of xenobiotics in STPs referred to a per capita per annum (cap*a) approach.

Compound	Source	Input to STP (g/cap*a)	Elimination (%)	Sorption to sludge (%)	Transformation (%)	Discharge (%)	Remarks
Triclosan	Household bactericide	0.25–1	87–95	~30	10 to bound residues ~5 to methyl triclosan	~5	Methyl triclosan may be a problem, as its bioaccumulation potential is higher than that of triclosan. Its elimination is triggered by biological processes.
HHCB	Fragrance	0.4–0.7	60–80	~50	~10 to the respective lactone	20–40	Elimination is due mostly to sorption to sludge; only a minor fraction is transformed by biological processes. The respective transformation product (HHCB-lactone) is highly mobile and persistent.
AHTN	Fragrance	0.09–0.15	60–70	~70	0	30–40	Elimination is due mostly to sorption to sludge.
TCPP	Flame-retardant	0.02–0.18	0–30	~70	0	70–100	TCPP cannot be treated via a per capita effectively, as there are probably point sources in space and time.
Musk xylene	Fragrance	0.004–0.01	70	2	Mostly to amino-musk xylene	~26	Musk xylene is mostly reduced to amino-musk xylene.
Musk ketone	Fragrance	0.006–0.015	52	10	0	47	Musk ketone is not eliminated well.
Bromocyclene	Pet-care insecticide		?	>90			Bromocyclene is accumulated in sludge, though the concentrations in the water phase under normal conditions are minute (0.7–12 ng L^{-1} in 1995 and 1996 [10]), as well as in surface waters (0.04–0.26 ng L^{-1}). Bromocyclene exhibits high bioaccumulation powers.

3. Reductive transformation, especially in the non-aerated parts of the activated sludge treatment. This might be especially relevant for dehalogenation processes. However, these processes are normally too slow to be performed within a few hours residence time of the activated sludge treatment. Thus, increasing the sludge and hydraulic retention time could help to perform this task, if this is possible. This might be a good strategy for halogenated compounds. However, for nitroaromates, the formed anilines are not wanted.

4. Oxidation procedures such as ozonization: Though these mechanisms can be quite efficient with regard to primary elimination of some compounds, it should be kept in mind that in most cases transformation products that might need to be controlled are formed. Thus, it might be a poor choice to oxidize unwanted but non-regulated chlorinated compounds to chlorophenols, which are highly toxic and highly regulated. It should also be kept in mind that some compounds needing controlled elimination such as the chlorinated organophosphates cannot be oxidized by the current means and installations.

5. Sorption to activated carbon: This is a quite powerful technique that works fine on compounds with a high K_{ow}. However, it might fail for compounds with a low K_{ow}, such as some pharmaceuticals and pesticides.

Probably the most efficient way to improve emissions of STPs is to combine two cost-effective technologies, e.g., ozonization and activated carbon treatment. Installations like this have been quite successful at eliminating xenobiotic compounds in drinking water treatment for decades.

4.2
Limnic Samples

Limnic ecosystems in industrialized countries are still heavily influenced by human activities. The big issues of the past such as huge point sources of toxic compounds in industrial areas (e.g., pesticide discharges at production plants) [268] have less importance in this century than in the past. On the other hand, more emissions of convenience chemicals (such as the bactericide triclosan), personal care compounds (such as synthetic fragrances), or medicinal compounds and flame-retardants have become relevant in production as well as economically and in everyday life. The same holds for emissions and contaminations of surface waters. From the mass balance point of view, at the moment fragrances and organophosphorus flame-retardants are among the most relevant compounds. The concentrations in surface waters are still in the hundreds of nanograms per liter, which means that transport from, e.g., the Ruhr to the Rhine is 100–500 kg/a. Most of this is transported from the rivers to the North Sea.

Some of these compounds seem to be as persistent as the PAHs or PCBs, which were the dominant issues of the past. These older compounds are still present in our environment, though their concentrations have decreased considerably. This is the case at least in the Ruhr megalopolis. Some of these "new" compounds can be transformed to other substances rapidly (such as HHCB-lactone or methyl triclosan); some of these compounds were identified during these studies, while others were already known. There is still very little knowledge about the toxicology issues of these transformation products. It is known that several of the compounds in question (such as the musk fragrances and triclosan) may be accumulated in fish and can undergo metabolic processes in these fish [12]. What we do not know is how this may affect the fish. It seems that the pollution connected with such compounds is currently much more widespread than that of the compounds initially regulated in documents such as the Water Framework Directive currently issued by the EU [4]. These compounds are emitted from basically everybody or every household and less so from economical activities such as industries and farming.

It will be interesting to note how the public will perceive the fact that, on the one hand, the issue of surface water contamination has improved because of de-industrialization and improvement in production processes, but, on the other hand, current solutions for personal security (flame-retardants, bactericides) and "beauty" (musk fragrances) lead to a completely different set of surface water contamination issues. Possibly, we need even better STPs to cope with these challenges. Alternatively, we could try to avoid persistent chemical in personal care and safety products, as in most cases there are indeed other solutions available. This will be as much an issue for consumers and citizens as for politicians and regulators.

4.3
Marine Samples

In marine samples it could be shown that basically all compounds present in limnic ecosystems are also present in marine ecosystems, e.g., the German Bight. This is not limited to the so-called POPs or special pesticides, nor are organochlorines the compounds with the highest concentrations. Examples of current pollution of the German Bight are given in Table 4.3 regarding a sample

Table 4.2 Regression functions (see Fig. 4.2) for diverse xenobiotics detected in the North Sea.

Compound	Slope (m)	B	r^2	Remarks
Nonylphenol	−0.561	21	0.84	Possibly transformed in the sea
MTB	−0.161	5.9	0.75	–
Xanthenone	−0.057	2.2	0.4	–
1-chloro-3-nitro-benzene	−0.037	1.4	0.5	–
2,5-dichloroaniline	−0.060	2.1	0.63	–

station over 50–100 km of the coast. In this compendium the classical organo-chlorines such as HCHs and PCBs exhibit rather low concentrations (0.5 ng L^{-1} and 1 ng L^{-1}, respectively) in comparison. Similar concentrations were determined for other compounds, whereas methyl thiophosphates, chlorinated butyl ethers, and triphenylphosphinoxide were found at concentrations one order of magnitude higher. Terbutylazine was detected at a concentration three orders of magnitude higher at that time. The triazine herbicides have been demonstrated to be within the concentrations that can trigger effects in the German Bight [269]. For other compounds the (eco)toxicology has not yet been well studied; thus, it is difficult to make predictions about whether these compounds will lead to effects. However, some of the discussed compounds are persistent, and some are highly bioaccumulating.

It seems that most of the compounds that we use reach not only the rivers but also the marine environment. To determine whether a compound is subject to transformation processes during its voyage from the sources in the rivers or the estuaries to the more central parts of the German Bight or even the open North Sea, sophisticated transport models can be used (compare Section 2.4.6). For a general impression, a salinity/analyte concentration plot may also give important information. In Fig. 4.2 these plots are performed for 2,5-dichloroaniline, methylthiobenzothiazole (MTB), nonylphenol, xanthenone, and 1-chloro-3-nitrobenzene.

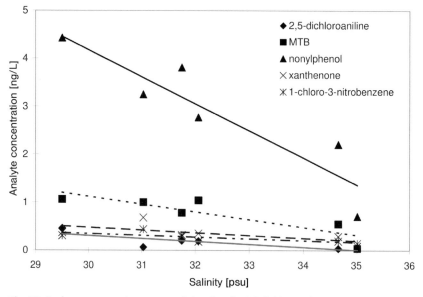

Fig. 4.2 Analyte concentration versus salinity plots for 2,5-dichloroaniline, MTB, nonylphenol, xanthenone, and 1-chloro-3-nitrobenzene derived from samples obtained from the Elbe estuary and the German Bight. Data from [60, 163, 164, 166, 168]. The same data as in Sections 2.1.4, 2.3, and 2.5.1 have been used.

Table 4.3 Comparison of concentrations of diverse compounds in the
German Bight at sample station 30.

Compound	Concentration (ng L^{-1})	Remarks	Reference
AHTN		See polycyclic musk fragrances	
Atrazine	42	Herbicide	[64]
a-HCH	0.5	Byproduct of lindane	[165]
Bis(4-chlorophenyl)-sulfone	1	Plasticizer	[168]
Caffeine	2	Soft drinks	[270]
Carbamazepine	2	Pharmaceutical	[270]
Chloropyridines	0.1–0.2		[270]
Chlorinated ethers (Cl$_4$BPE)	30	Byproducts from, e.g., epichlorohydrin synthesis	[35]
DEET (*N,N*-diethyl-3-toluamide)	0.6	Midge repellent	[270]
2,5-Dichloroaniline	0.4	Degradation of pesticides, pigments etc.	[164]
HHCB		See polycyclic musk fragrances	
Methyl thiophosphates	1–8	Chemical waste from pesticide production	[63]
Methyl-thio-benzothiazole	0.4–1.4	Degradation of tires, rubber, and TCMTBT	[60]
Musk ketone	<0.02–0.12	Fragrance	[62]
Musk xylene	<0.03–0.08	Fragrance	[62]
Nitrobenzenes	<0.05–1.0	Chemical waste	[62, 166]
Nonylphenols	2.5	Degradation products of detergents	[168]
PCBs	~1	Electrical insulators	[271]
Polycyclic musk fragrances such as HHCB and AHTN	0.2–0.6	Fragrances for shampoos, washing powder	[39]
Propyphenanzone	0.6	Pharmaceutical	[270]
TCPP	1–8	Flame-retardant	[272]
Terbutylazine	360	Herbicide	[64]
Triphenyl phosphinoxide	53		[270]
Xanthen-9-one	0.21-0.37	Coal, automobile exhaust	[163]

TCMTBT = thiocyanate methylthiobenzothiazole; PCBs = polychlorinated biphenyls.

It can be shown that compounds such as MTB are distributed conservatively in the German Bight. Thus, only stratification/dilution processes take place. Compounds such as HHCB are sorbed to sediment, are volatile, or are degraded to some extent. However, in the more recent datasets, HHCB also seems to be more or less stable in the marine environment.

From concentration versus salinity plots (Fig. 4.2), trend lines can be obtained in the form of a linear dependency:

$$c = m \times salinity + b$$

where c=concentration, m=slope (derives from degradation and dilution), and b=concentration at salinity=0 psu.

A steep slope indicates more rapid phase transitions or transformation processes, while a low slope indicates dilution. As can be determined in Table 4.2, nonylphenol shows a much steeper slope than all other compounds in this experiment. Thus, this compound is probably not simply transported conservatively into the German Bight (diluted) but rather undergoes some kind of transformation process. On the other hand, other compounds (MTB, dichloroaniline, etc.) are probably just diluted into the seawater. Relatively low correlations for xanthenone and chloronitrobenzene are probably the result of local processes. In this case the samples represent an area of more than 200×200 km and the water masses need 6 to 12 months to pass from the mouth of the river (Elbe) to the northern boundaries of the experimental area (German Bight).

4.4
Conclusions

Only a few of the organic micropollutants studied can be eliminated effectively and repeatedly in sewage treatment processes. Only triclosan is more or less quantitatively removed from the wastewater stream, though some doubts remain about whether it is transformed (metabolized) in the STP. The polycyclic musk fragrances HHCB and AHTN exhibit some sorption to sludge but no total removal. The nitroaromatic musks exhibit diverse reactions. While musk ketone is not eliminated at all and is present in the respective river at the resulting concentrations, musk xylene, for some reason, is metabolized faster in the STP to form the respective anilines. On the other hand, the chlorinated organophosphates are not eliminated at all, and they are found quantitatively in surface waters.

In the end, all of the compounds released from the STPs into the rivers were also found in marine waters. In none of these cases does degradation in the environment result in a situation such that the respective compound is significantly reduced in the environment of the surface waters. On the other hand, sorption to sediment may be relevant; unfortunately, this is currently detectable in only a few cases. Thus, it seems that a steady state with sorption/desorption may be more relevant for the medium lipophilic compounds discussed in this book.

5
Summary

In this study the fate of xenobiotics has been studied. The focus has been on personal care compounds such as the polycyclic musk fragrances HHCB, AHTN, and OTNE; flame-retardants such as TCPP; and bactericides such as triclosan. All of these compounds might be considered "lifestyle chemicals." The main subjects have been emission, passage of sewage treatment facilities, possible phase transformation, and chemical and biochemical conversion to metabolites or other transformation products, together with loads in surface waters, including marine waters. An overview of the fate of these compounds is given in Table 5.1.

With respect to the sewage treatment process, total balances including influent, effluent, and sludge were performed for the first time for these compounds. Also, multistage process studies (including primary, secondary, and tertiary treatment) were performed (Sections 2.1.1 and 2.4.2). Some attention has also been drawn to consumer protection as well as soil contamination with antibiotics used in agriculture.

For some compounds it was possible to add data for overall assessment or to gain new insights, e.g., for bromocyclene, without describing the whole aquatic pathway as for the other compounds.

To perform these tasks, methods to determine compounds in low concentrations were established. The methods were operational in the range of micrograms per liter in wastewater, nanograms per liter in limnic surface waters, and even in picograms per liter in seawater. In most cases GC-MS was the final method of determining the compounds in question, but in some cases HPLC-MS/MS was applied. For several methods, special pitfalls such as matrix effects in HPLC-MS/MS were determined as a result of careful method validation procedures. Additionally, steps to overcome the respective problems were found and published.

5.1
Polycyclic Musks AHTN, HHCB, HHCB-lactone, and OTNE

Though the detergent industry states that all but one manufacturer has phased out polycyclic musk fragrances and nitroaromatic musk fragrances from washing powders, AHTN, HHCB, and the transformation product HHCB-lactone

Personal Care Compounds in the Environment: Pathways, Fate, and Methods for Determination. Kai Bester
Copyright © 2007 WILEY-VCH Verlag GmbH & Co. KGaA, Weinheim
ISBN: 978-3-527-31567-3

Table 5.1 Overview on the fate of diverse compounds in STPs, surface waters, and marine ecosystems.

Compound	Sewage treatment plants (as derived from STPs A, B and C)				Surface water (as derived from the Ruhr, Rhine, and Lenne)		Marine ecosystems	Remarks
	Introduction per capita per annum (g)	Elimination rate[a] (%)	Elimination mechanism	Discharge per capita per annum (g)	Concentrations (ng L^{-1})	Half-life[b] (d)	Concentrations (ng L^{-1})	
AHTN	0.09–0.15	57–78	Sorption to sludge: 100%	0.03–0.05	10	15	0.08–2.6	AHTN was removed from washing powders by most manufacturers in Germany in the years 1994–2004
Bromo-cyclene	~0.0003	Probably large	Sorption to sludge; biodegradation	n.d.	0.05–0.1 ng L^{-1} in 1995 in northern Germany; 2003: <1 ng L^{-1}	n.d.	n.d.	Application and production of bromocyclene were officially phased out in 1994 in Germany
HHCB	0.40–0.66	54–73	Sorption to sludge; biotransformation to HHCB-lactone (~10% total)	0.17–0.18	50	67	0.09–4.8	HHCB was removed from washing powders by most manufacturers in Germany in the years 1994–2004
HHCB-lactone	0.03–0.06	−6 to −60[a]	Generation from HHCB	0.05–0.09	30	~40	0.2–1	

Musk ketone	0.006–0.015	12–51	Probably transformation to musk ketone amine; sorption to sludge is negligible (<5%)	~0.01	~1.5	33	<0.03–0.12 [62]
Musk xylene	0.004–0.001	73–95	Transformation to musk xylene amine; sorption to sludge is negligible (<5%)	0.0001–0.0032	<1	n.d.	<0.02–0.08 [62]	Musk xylene has been removed from the market in Germany, but it will still be present in other industrialized countries
OTNE	0.86	~73	Sorption to sludge	0.25	~150	n.d.	n.d.
TCPP	0.13–0.65	−7.4[a] to 21	None	0.09–0.64	150	98	1–8
Triclosan	0.25–1.04	87–96	Sorption to sludge: 30%; bound residues >10%; transformation to methyl triclosan ~1% residue: unknown	0.01–0.10	5	11	Detectable only near STP effluents
Methyl triclosan	–		Generated from triclosan in the STP	0.007	1.5	11	n.d.

a) Negative numbers imply generation during the activated sludge process.
b) Half-life refers to in-river elimination obtained by comparison of concentrations at various points in the river with known hydrodynamic retention, i.e., runtime of the water from one point to the other. In this case Lake Kemnaden and Lake Mülheim (stations 61 and 66) are used. The total uncertainty of this approach is about 30% of the concentration. Main parameters contributing to this uncertainty are the emissions and the determination of concentration.

are still present in wastewater, STP effluent, sewage sludge, and surface waters. In this study the transformation of HHCB to HHCB-lactone has been demonstrated for the first time with a balancing experiment including sludge as well as influent and effluent of a sewage treatment plant. It has also been shown that this transformation is performed by microorganisms and not by chemical oxidation or by photo-oxidation. However, it was demonstrated that the main mechanism of elimination of HHCB (54–73%; Table 5.1) and AHTN (57–78%) from wastewater is sorption to sludge. In the surface waters of northern Germany, i.e., the Ruhr, degradation of these compounds is hardly observable, and estimation of half-lives is 15, 40, and 67 d for AHTN, HHCB-lactone, and HHCB, respectively, in the Ruhr. Compared to the relatively short hydrodynamic residence times in rivers such as the Ruhr (i.e., in eight days the whole Ruhr megalopolis is passed), in-river elimination is not effective in reducing polycyclic musk compounds. No assessment of HHCB-lactone in STP balances and surface waters was published before, and enantioselective degradation of HHCB-lactone was not determined in STPs previously.

Both HHCB and AHTN were detected in the marine ecosystem of the German Bight of the North Sea. Though the concentrations were low in comparison, and though some elimination of both compounds was observed in the estuary, the presence of both compounds does not comply with the zero-emission criterion of OSPAR to reduce the load of xenobiotics in the North Sea.

The voluntary phase-out of these compounds from washing powders has not yet been effective in removing them from the environment totally. However, it is likely that other sources of these compounds contribute to their loads in the environment.

OTNE is not exactly a polycyclic musk fragrance, as its sensory effect is different. It had been observed in only one study considering elimination prior to this study. In this study, OTNE was observed at higher concentrations than any other fragrance that was studied. Its emission rate is about 1 g per capita annually. It is eliminated by 66% in STPs; most of this elimination is due to sorption to sludge, as most of the eliminated OTNE was determined in the sludge. This fragrance was also detected in surface waters and is persistent under ambient conditions. No sludge or surface water data or emission scenarios were available in previous studies, and OTNE had not yet been studied in marine ecosystems.

5.2
Flame-retardants

The focus in this study was on organophosphate flame-retardants such as TCPP but not on organobromine compounds, because organophosphates are water-soluble and thus highly mobile. About 95% of TCPP is used in rigid polyurethane foam plates, which are used in thermal building insulation. In contrast to the polycyclic musk fragrances as well as triclosan, organophosphates have been determined with a high variability in wastewater. It was concluded that the main source of TCPP might be construction activities, as all other possibilities either have been ruled out or should lead to continuous emissions, such as leaching from buildings. TCPP was not eliminated significantly in any of the studied STPs. The concentrations in STP influents (500–4500 ng L^{-1}) and effluents (400–4500 ng L^{-1}) are considerable. Additionally, the per-capita emission rates in the respective STPs differed by nearly an order of magnitude (Table 5.1). No sludge data on this compound were available before this study, nor were any data-based assessments of emissions.

The concentrations in the Ruhr were also high in comparison: 150 ng L^{-1} after passing the first major cities. TCPP is not eliminated in the river, as the estimated in-river half-life is >90 d. Basically, the Ruhr transports 300 kg of this compound annually towards the Rhine. The concentrations in the Rhine and the Lippe are similar to those in the Ruhr. No transport assessments were available before this study was performed. In the river Rhine ten times more of TCPP is transported in the river Ruhr, which reflects the size of these two rivers. These amounts are transported to The Netherlands and the North Sea.

Considerable concentrations (i.e., 1–8 ng L^{-1}) of TCPP were determined in marine samples. A comparison of inventories in the German Bight with the annual inputs via large rivers supports the hypothesis that this compound is distributed by dilution in the German Bight and that no means of degradation are relevant in the marine environment. TCPP does not comply with the OSPAR criterion of zero emissions to the North Sea. TCPP is currently undergoing a re-evaluation under the high-production chemicals legislation of the EU. Though the toxicity of TCPP is not very high, TCPP will probably be evaluated under the Water Framework Directive in the near future.

5.3
Endocrine Disrupters

Endocrine disrupters such as nonylphenol and bisphenol A have been studied in a multitude of other studies in limnic systems as well as wastewater treatment. Thus, the focus of our study was to determine the endocrine disrupter nonylphenol in seawater. It was demonstrated that this compound indeed is present in the North Sea at concentrations similar to other pollutants (Table 4.3). It was thus demonstrated that endocrine disrupters such as nonylphenols do reach open marine ecosystems. It was thus established that nonylphenols do not comply with the OSPAR regulations or the Water Framework Directive.

5.4
Triclosan and Methyl Triclosan

The use of triclosan as a domestic bactericide has led to high loads in sewage waters. This compound is used with some regional variability in Germany, but usage rates range from 0.25 g to 1 g per capita annually. In contrast to the fragrances, triclosan is eliminated effectively (87–96%; Table 5.1) in today's sewage treatment plants. The majority of the triclosan is sorbed to the sludge, while some bound residues are formed in the sludge as well. Data from this study suggested that this is at least partially a biologically induced process. The known metabolite methyl triclosan is produced in STPs, but this is only a small fraction of the total ($\sim 1\%$). On the other hand, this transformation product is relevant in STPs effluents. The concentrations of methyl triclosan in the effluents are about 20–30% of that of triclosan.

This ratio is stable in the Ruhr; thus, no strong indications for in-river elimination were found. Tentative half-lives of 11 d were determined for both compounds, which is compared to hydrodynamic retention in this part of the Ruhr of about 6 d. Photodegradation as discussed for Swiss lakes was not relevant for this ecosystem.

Triclosan is currently being assessed under the Persistent Organic Pollutant (POP) convention. Because of its high accumulation power, methyl triclosan may well prove to be more relevant than triclosan itself. However, the main path of human exposition for compounds such as triclosan is household usage.

References

1 Carson, R.: Silent spring, Penguin Books, Harmonsworth, England, 1988
2 Colborn, T., Dumnanoski, D., Peterson-Myers J. P.: Our stolen future: are we threatening our fertility, intelligence, and survival? Plume Books, New York, 1997
3 Jobling, S., Monique Nolan, M., Tyler, C.R., Brighty, G., Sumpter J. P.: Widespread Sexual Disruption in Wild Fish, *Environ. Sci. Technol.*, 32, 2498–2506 (1998)
4 European Commission: EC 2000/60, Water Framework Directive, Official Journal, L 327, 22.12.2000, P1
5 Bester, K.: Chiral analysis for environmental applications, *Anal. Bioanal. Chem.*, 376, 302–304 (2003)
6 OSPAR: Ministerial Declaration of the fourth International Conference on the protection of the North Sea, Esbjerg, Denmark, 1995 "The Ministers AGREE that the objective is to ensure a sustainable, sound and healthy North Sea ecosystem. The guiding principle for achieving this objective is the precautionary principle. This implies the prevention of the pollution of the North Sea by continuously reducing discharges, emissions and losses of hazardous substances thereby moving towards the target of their cessation within one generation (25 years) with the ultimate aim of concentrations in the environment near background values for naturally occurring substances and close to zero concentrations for man-made synthetic substances."
7 Hühnerfuss, H., Faller, J., Kallenborn, R., König, W.A., Ludwig, P., Pfaffenberger, B., Oehme, M., Rimkus, G.: Enantioselective and nonenantioselective degradation of organic pollutants in the marine ecosystem, *Chirality*, 5, 393 (1993)
8 Kallenborn, R., Hühnerfuss, H.: Chiral Environmental Pollutants, Springer, Berlin, 2001
9 Kallenborn, R., Oehme, M., Vetter, W., Parlar, H.: Enantiomer selective separation of toxaphene congeners isolated from seal blubber and obtained by synthesis, *Chemosphere*, 28, 89 (1994)
10 Bethan, B., Bester, K., Hühnerfuss, H., Rimkus, G.: Bromocyclen contamination of surface water, wastewater and fish from Northern Germany and gas chromatographic chiral separation, *Chemosphere*, 34, 2271(1997)

11 Buser, H.R., Müller, M.D.: Environmental behaviour of acetamide pesticide stereoisomers. 1. Stereo- and Enantioselective determination using chiral high resolution gas chromatography and chiral high performance liquid chromatography, *Environ. Sci. Technol.* 29, 2023 (1995)

12 Gatermann, R., Biselli, S., Hühnerfuss, H., Rimkus, G.G., Franke, S., Hecker, M., Kallenborn, R., Karbe, L., Konig, W.A.: Synthetic musks in the environment. Part 2: Enantioselective transformation of the polycyclic musk fragrances HHCB, AHTN, AHDI, and ATII in freshwater fish, *Arch. Environ. Contam. Toxicol.*, 42, 447 (2002)

13 Ward, T.J.: Chiral separations, *Anal. Chem.* 74, 2863 (2002)

14 Vetter, W., Smalling, K.L., Maruya, K.A.: Interpreting nonracemic ratios of chiral organochlorines using naturally contaminated fish, *Environ. Sci. Technol.* 35, 4444 (2001)

15 Monkiedje, A., Spiteller, M., Bester, K.: Degradation behaviour and effects on microbial biomass of racemic mixture and enantiomeric forms of the fungicide metalaxyl in two soils, *Environ. Sci. Technol.* 37, 707–712 (2003)

16 Hardt, I., Wolf, O., Gehrcke, B., Hochmuth, D.H., Pfaffenberger, B., Hühnerfuss, H., König, W.: Gaschromatographic enantiomer separation of agrochemicals and polychlorinated biphenyls (PCBs) using modified cyclodextrins, *J. High Resolut. Chromatogr.*, 17, 859–864 (1994)

17 Aigner, E.J., Leone, A.D., Falconer, R.L.: Concentrations and enantiomeric ratios of organochlorine pesticides in soil from the US Corn Belt, *Environ. Sci. Technol.*, 32, 1162–1168 (1998)

18 Bucheli, T., Müller, S.R., Voegelin, A., Schwarzenbach, R.P.: Bituminous roof sealing membranes as major sources of the herbicide (R,S)-Mecoprop in roof runoff waters: potential contamination of groundwater and surface waters, *Environ. Sci. Technol.*, 32, 3465 (1998)

19 Wong, C.S., Lau, F., Clark, M., Mabury, S.A., Muir, D.C.G.: Rainbow trout (Oncorhynchus mykiss) can eliminate chiral organochlorine compounds enantioselectively, *Environ. Sci. Technol.* 36, 1257–1262 (2002)

20 Karlsson, H., Oehme, M., Skopp, S., Burkow, I.C.: Enantiomer ratios of chlordane congeners are gender specific in cod (*Gadus morhua*) from the Barents Sea, *Environ. Sci. Technol.* 34, 2126–2130 (2000)

21 Ridal, J.J., Bidleman, T.F., Kerman, B.R., Fox, M.E., Strachan, W.M.J.: Enantiomers of a-Hexachlorocyclohexane as tracers of air-water gas exchange in lake Ontario, *Environ. Sci. Technol.* 31, 1940 (1997)

22 Dsikowitzky, L., Schwarzenbauer, J., Littke, R.: Distribution of polycyclic musks in water and particulate matter of the Lippe river (Germany), *Organic Geochemistry*, 33, 1747–1758 (2002)

23 Eschke, H.D., Dibowski, H.J., Traud, J.: Untersuchungen zum Vorkommen polycyclischer Moschus-Duftstoffe in verschiedenen Umweltkompartimenten, *UWSF-Z. Umweltchem. Ökotox.*, 7, 131–138 (1995)

24 Eschke, H.D., Traud, J., Dibowski, H.J.: Untersuchungen zum Vorkommen polycyclischer Moschusduftstoffe in verschiedenen Umweltkompartimenten-

Nachweis und Analytik mit GC-MS in Oberflächen-, Abwässern und Fischen (1. Mitteilung), *UWSF-Z. Umweltchem. Ökotox.*, 6, 183–189 (1994)

25 Simonich, S. L., Begley, W. M., Debaere, G., Eckhoff, W. S.: Trace analysis of fragrance materials in wastewater and treated wastewater, *Environ. Sci. Technol.*, 34, 959–965 (2000)

26 Simonich, S. L., Federle, T. W., Eckhoff, W. S., Rottiers, A., Webb, S., Sabaliunas, D., De Wolf, W.: Removal of fragrance materials during US and European wastewater treatment, *Environ. Sci. Technol.*, 36, 2839–2847 (2002)

27 Paxeus, N.: Organic pollutants in the effluents of large wastewater treatment plants in Sweden, *Water Res.*, 30, 1115–1122 (1996)

28 Rimkus, G. G.: Synthetik musk fragrances in the environment, Springer, Heidelberg, 2004

29 Bester, K., Hühnerfuss, H.: Triazine herbicide concentrations in the German Wadden Sea, *Chemosphere*, 32, 1919–1928 (1996)

30 Gatermann, R., Hellou, J., Hühnerfuss, H., Rimkus, G., Zitko, V.: Polycyclic and nitro musks in the environment: A comparison between Canadian and European aquatic biota, *Chemosphere*, 39, 1571–1571 (1999)

31 Gatermann, R., Biselli, S., Hühnerfuss, H., Rimkus, G. G., Hecker, M., Karbe, L.: Synthetic musks in the environment. Part 1: Species-dependent bioaccumulation of polycyclic and nitro musk fragrances in freshwater fish and mussels, *Arch. Environ. Contam. Toxicol.*, 42, 437–446 (2002)

32 Rimkus, G., Wolf, M.: Polycyclic Musk fragrances in human adipose tissue and human milk, *Chemosphere*, 33, 2033–2043 (1996)

33 Seinen, W., Lemmen, J. C., Pieters, R. H. H., Verbruggen, E. M. J., van der Burg, B.: AHTN and HHCB show weak estrogenic – but no uterotrophic activity, *Toxicol. Let.*, 111, 161–168 (1999)

34 Kallenborn, R., Gatermann, R., Nygard, T., Knutzen, J., Schlabach, M.: Synthetic musks in Norwegian marine fish samples collected in the vicinity of densely populated areas, *Fresenius Environ. Bull.*, 10, 832–842 (2001)

35 Franke, S., Meyer, C., Heinzel, N., Gatermann, R., Hühnerfuss, H., Rimkus, G., Konig, W. A., Francke, W.: Enantiomeric composition of the polycyclic musks HHCB and AHTN in different aquatic species, *Chirality*, 11, 795–801 (1999)

36 Ricking, M., Schwarzbauer, J., Hellou, J., Svenson, A., Zitko, V.: Polycyclic aromatic musk compounds in sewage treatment plant effluents of Canada and Sweden – first results, *Mar. Pollut. Bull.*, 46, 410–417 (2003)

37 Heberer, T.: Occurance, fate, and assessment of polycyclic musk residues in the aquatic environment of urban areas – a review, *Acta Hydrochim. Hydrobiol.*, 30, 227–243 (2002)

38 Bester, K.: Triclosan in sewage plants -balances and monitoring data-, *Water Res.*, 37, 3891–3896 (2003)

39 Bester, K., Hühnerfuss, H., Lange, W., Rimkus, G. G., Theobald, N.: Results of non target screening of lipophilic organic pollutants in the German Bight II: Polycyclic musk fragrances, *Water Res.*, 32, 1857–1863 (1998)

40 Balk, F., Ford, R. A.: Environmental risk assessment for the polycyclic musks AHTN and HHCB in the EU I fate and exposure assessment, *Environ. Toxicol. Let.*, 111, 57–79 (1999)

41 Herren, D., Berset, J. D.: Nitro musks, nitro musk amino metabolites and polycyclic musks in sewage sludges – Quantitative determination by HRGC-ion-trap-MS/MS and mass spectral characterization of the amino metabolites, *Chemosphere*, 40, 565–574 (2000)

42 Lee, H. B., Peart, T. E., Sarafin K.: Occurance of polycyclic and nitro musk compounds in Canadian sludge and wastewater samples, *Water Qual. Res. J. Canada*, 38, 683–702 (2003)

43 Kupper, T., Berset, J. D., Etter-Holzer, R., Furrer, R., Tarradellas, J.: Concentration and specific loads of polycyclic musks in sewage sludge originating from a monitoring network in Switzerland, *Chemosphere*, 54, 1111–1120 (2004)

44 Rimkus, G. G.: Polycyclic musk fragrances in the aquatic environment, *Toxicol. Lett.*, 111, 37–56 (1999)

45 Dsikowitzky, L., Schwarzenbauer, J., Littke, R.: Distribution of polycyclic musks in water and particulate matter of the Lippe river (Germany), *Org. Geochem.*, 33, 1747–1758 (2002)

46 Ternes, T. A., Stuber, J., Herrmann, N., McDowell. D., Ried, A., Kampmann, M., Teiser, B.: Ozonation: a tool for removal of pharmaceuticals, contrast media and musk fragrances from wastewater? *Water Res.*, 37 1976–1982 (2003)

47 Bester, K.: Polycyclic musks in the Ruhr catchment area – Transports, discharges of wastewater, and transformations of HHCB, AHTN, and HHCB-lactone, *J. Environ. Monitor.*, 7, 43–51 (2005)

48 Buerge, I. J., Buser, H. R., Muller, M. D., Poiger, T.: Behavior of the polycyclic musks HHCB and AHTN in lakes, two potential anthropogenic markers for domestic wastewater in surface waters, *Environ. Sci. Technol.*, 37, 5636–5644 (2003)

49 Bester, K.: Polycyclic musks in the Ruhr catchment area – Transports, discharges of wastewater, and transformations of HHCB, AHTN, and HHCB-lactone, *J. Environ. Monitor.*, 7, 43–51 (2005)

50 Heberer, T., Gramer, S., Stan, H. J.: Occurance and distribution of organic contaminants in the aquatic system in Berlin. Part III Determination of synthetic musks in Berlin surface water applying solid phase microextraction (SPME) and gas chromatography-mass spectrometry, *Acta Hydrochim. Hydrobiol.*, 27, 150–156 (1999)

51 Meyer, C.: Screening, identification and enantioselective analysis of organic compounds in surface waters (in German), PhD-thesis, University Hamburg (2001)

52 Internationale Kommission zum Schutz des Rheines (IKSR): Zahlentafeln der physikalisch-chemischen Untersuchungen des Rheinwassers und des Schwebstoffes 1994, Internationale Kommission zum Schutz des Rheins (IKSR), Koblenz, 1996

53 Federle, T. W., Itrich, N. R., Lee D. M., Langworthy D.: Recent advances in the environmental fate of fragrance ingredients, SETAC, 2002, Vienna

54 Lagois, U.: Vorkommen von synthetischen Nitromoschusverbindungen in Gewässern. (Occurrence of synthetic nitromusk fragrance compounds in surface waters), *gwf* – *Wasser Abwasser*, 137, 154–155 (1996)

55 Franke, S., Hildebrandt, S., Schwarzbauer, J., Link, M., Francke, W.: Organic compounds as contaminants of the Elbe river and its tributaries part II, GC-MS screening for contaminants of the Elbe water, *Fresenius J. Anal. Chem.*, 353, 39–49 (1995)

56 Rimkus, G.G., Brunn, H.: Synthetische Moschusduftstoffe – Anwendung, Anreicherung in der Umwelt und Toxikologie – Teil 1 Herstellung, Anwendung, Vorkommen in Lebensmitteln, Aufnahme durch den Menschen. (Synthetic musk compounds – application, environmental accumulation, and toxicity, part 1 production, application, presence in food, uptake by humans), *Ernährungs – Umschau*, 43, 442–449 (1996)

57 Barbetta, L., Trowbridge, T., Eldib, I.A.: Musk aroma chemical industry, *Perfumer and Flavorist*, 13, 60–618 (1988)

58 Brunn, H., Rimkus, G.: Synthetische Moschusduftstoffe – Anwendung, Anreicherung in der Umwelt und Toxikologie – Teil 2 Toxikologie der synthetischen Moschusduftstoffe und Schluβfolgerungen. (Synthetic musk compounds – application, environmental accumulation, and toxicity, part 2 toxicology and conclusions), *Ernährungs – Umschau*, 44, 4–9 (1997)

59 Müller, S., Schmid, P., Schlatter, C.: Occurrence of nitro and non-nitro benzoid musk compounds in human adipose tissue, *Chemosphere*, 33, 17–28 (1996)

60 Bester, K., Hühnerfuss, H., Lange, W., Theobald N.: Results of non target screening of lipophilic organic pollutants in the German Bight I: Benzothiazoles, *Sci. Total Environ.*, 207, 111–118 (1997)

61 Theobald, N., Gaul, H., Ziebarth, U.: Verteilung von organischen Schadstoffen in der Nordsee und angrenzenden Seegebieten. (Distribution of organic pollutants in the North Sea and adjoining areas), *Dt. Hydrogr. Z., Suppl.* 6, 81–93 (1996)

62 Gatermann, R., Hühnerfuss, H., Rimkus, G., Wolf, M., Franke, S.: The distribution of nitrobenzene and other nitroaromatic compounds in the North Sea, *Mar. Pollut. Bull.*, 30, 221–227 (1995)

63 Gatermann, R., Bester, K., Franke, S., Hühnerfuss, H.: The distribution of O,O,O-trimethylthiophosphate and O,O,S-trimethyldithiophosphate in the North Sea, *Chemosphere*, 32, 1907–1918 (1996)

64 Bester, K., Hühnerfuss, H. Triazines in the Baltic and North Sea, *Mar. Pollut. Bull.*, 26, 423–427 and 657–658 (1993)

65 Eschke, H.-D., Dibowski, H.-J., Traud, J.: Nachweis und Quantifizierung von polycyclischen Moschus-Duftstoffen mittels Ion-Trap GC/MS/MS in Humanfett und Muttermilch. (Detection and quantification of polycyclic musk fragrances with ion-trap GC/MS/MS in human adipose tissue and human milk), *Dtsch. Lebensm.-Rdsch.*, 91, 375–379 (1995b)

66 Spencer, P.S., Sterman, A.B., Horopian, D.S., Foulds, M.M.: Neurotoxic fragrance produces ceroid and Myelin Disease, *Science*, 204, 633–635 (1979)

67 Gautschi, M., Bajgrowicz, J. A., Kraft, P.: Fragrance Chemistry – milestones and perspectives, *Chimia*, 55, 379–387 (2001)

68 Bester, K.: Retention characteristics and balance assessment for two poly-cyclic musk fragrances (HHCB and AHTN) in a typical German sewage treatment plant, *Chemosphere*, 57, 863–870 (2004)

69 Aschmann, S. M., Arey, J., Atkinson, R., Simonich, S. L.: Atmospheric life-times and fates of selected fragrance materials and volatile model com-pounds, *Environ. Sci. Technol.*, 35, 3595–3600, 2001

70 Difrancesco, A. M., Chui, P. C., Standley, L. J., Allen, H. E., Salvito, D. T.: Dissipation of fragrance materials in sludge-amended soils, *Environ. Sci. Technol.*, 38, 194–201 (2004)

71 Nussbauer, C., Frater, G., Kraft, P.: (±)-1-[(1R*,2R*,8aS*)-1,2,3,5,6,7,8,8a-Octa-hydro-1,2,8,8-tertamethylnaphthale-2-yl]ethan-1one: isolation and stereoselec-tive synthesis of a powerful minor constituent of the perfumery synthetic Iso E Super, *Helvitica Chimica Acta*, 83, 1016–1024 (1999)

72 Mersch-Sundermann, V., Reihardt, A., Emig, M.: Examination of mutageni-city, genotoxicity, and cogenotoxicity of nitro musks in the environment, *Zbl. Hyg.*, 198, 429–442 (1996)

73 Adolfsson-Erici, M., Pettersson, M., Parkkonen J., Sturve J.: Triclosan, a com-monly used bactericide found in human milk and in the aquatic environ-ment in Sweden, *Chemosphere*, 46, 1485–9 (2002)

74 Lopez-Avila, V., Hites, R. A.: Organic compounds in an industrial wastewater. Their transport into sediments, *Environ. Sci. Technol.*, 14, 1382–90 (1980)

75 McAvoy, D. C., Schatowitz B., Jacob M., Hauk, A., Eckhoff, W. S., Measure-ment of Triclosan in wastewater treatment systems, *Environ. Toxicol. Chem.*, 21, 1323–1329 (2002)

76 Lindström, A., Buerge, I. J., Poiger, T., Bergqvist, P.-A., Muller, M. D., Buser, H.-R.: Occurrence and environmental behavior of the bactericide Triclosan and its methyl derivative in surface waters and in wastewater, *Environ. Sci. Technol.*, 36, 2322–9 (2002)

77 Hale, R. C., Smith, C. L.: A multi-residue approach for trace organic pollu-tants: application to effluents and associated aquatic sediments and biota from the southern Chesapeake Bay drainage basin 1985–1992, *Int. J. Environ. Anal. Chem.*, 64, 21–33 (1996)

78 Orvos, D. R., Versteeg, D. J., Inauen, J., Capdevielle, M., Rothenstein, A.: Aquatic toxicity of Triclosan, *Environ. Toxicol. Chem.*, 21, 1338–1349 (2002)

79 Federle, T. W., Kaiser, S. K., Nuck, B. A.: Fate and effects of Triclosan in acti-vated sludge, *Environ. Toxicol Chem.*, 21, 1330–1337 (2002)

80 Singer, H., Müller, S., Tixier, C., Pillonel, L.: Triclosan: Occurrence and Fate of a Widely Used Biocide in the Aquatic Environment: Field meassurements in wastewater treatment plants, surface waters, and lake sediments, *Environ. Sci. Toxicol.*, 36, 4998–5004 (2002)

81 Bester, K.: Fate of Triclosan and Triclosan-Methyl in sewage treatment plants and surface waters, Arch. *Environ. Contam. Toxicol.*, 49, 9–18 (2005)

82 Ruhrverband (2004) Anhaltswerte Fliesdatentabelle für die Ruhr (Niedrigwasser) (Waterflow of the river Ruhr)

83 Morrall, D., McAvoy, D., Schatowitz, B., Inauen, J., Jacob, M., Hauk, A., Eckhoff, W.: A field study of Triclosan loss rates in river water (Cibolo Creek, TX), *Chemosphere*, 54, 653–660 (2004)

84 Wind, T., Werner, U., Jacob, M., Hauk, A.: Environmental concentrations of boron, LAS, EDTA, NTA and Triclosan simulated with GREAT-ER in the river Itter, *Chemosphere*, 54, 1135–1144 (2004)

85 Sabaliunas, D., Webb, S. F., Hauk, A., Jacob, M., Eckhoff, W. S.: Environmental fate of Triclosan in the River Aire Basin, UK, *Water Res.*, 37, 3145–3154 (2003)

86 Tixier, C., Singer, H., Canonica, S., Müller, S.: Phototransformation of Triclosan in surface waters: A Relevant elimination process for this widely used biocide-laboratory studies, field measurements, *Environ. Sci Toxicol.*, 36, 3482–3489 (2002)

87 Balmer, M. E., Poiger, T., Droz, C., Romanin, K., Bergqvist, P.-A., Müller, M. D., Buser, H.: Occurrence of Methyl Triclosan, a Transformation Product of the Bactericide Triclosan, in Fish from Various Lakes in Switzerland, *Environ. Sci. Technol.*, 38, 390–395 (2004)

88 Balmer, M. E., Buser, H. R., Müller, M. D., Poiger T.: Occurance of Some Organic UV filters in wastewater, in surface waters and in fish from Swiss Lakes, *Environ. Sci. Technol.* 39, 953–962 (2005)

89 Schlumpf, M., Cotton, B., Conscience, M., Haller, V., Steinmann, B., Lichtensteiger, W.: In vitro and in vivo estrogenicity of UV Screens, *Environ. Health Perspect.*, 109, 239–244, 2001

90 European Commission: Towards the establishment of a priority list of substances for further evaluation of of their role in endocrine disruption, final report DG Environment, M0355008/1786Q, Delft, The netherlands, 2000

91 Nagtegaal, M., Ternes, T., Baumann, W., Nagel, R.: UV-Filtersubstanzen in Wasser und Fischen, *Z. Umweltchem. Ökotoxikol.*, 9, 79–81 (1997)

92 Poiger, T., Buser, H. D., Balmer, M. E, Bergqvist, P. A., Müller, M. D.: Occurance of UV filter compounds from sunscreens in surface waters:regional mass balance in two lakes, *Chemosphere*, 55, 951–963 (2004)

93 Buser, H. D., Müller, M. D., Balmer, M. E., Poiger, T., Buerge, I. J.: Stereoisomer composition of the Chiral UV Filter 4-Methylbenzylidene Camphor in Environmental Samples, *Environ. Sci. Technol.*, 39, 3013–3019 (2005)

94 Leisewitz, A., Kruse, H., Schramm, E.: Erarbeitung von Bewertungsgrundlagen zur Substitution umweltrelevanter Flammschutzmittel, *Umweltbundesamt Berichte* 25/01 (in German) (2000)

95 IAL market report: The European flame-retardant chemical industry 1998, IAL consultants, London (1999)

96 CEFIC private communication. To the European Flame-retardants Association (2002)

97 Carlsson, H., Nilsson, U., Becker, G., Östman, C.: Organophosphate ester flame-retardants and plasticisers in the indoor environment: Analytical methodology and occurance, *Environ. Sci. Technol.*, 31, 2931–2936 (1997)

98 Kemmlein, S., Hahn, O., Jann, O.: Emissions of organophosphate and brominated flame-retardants from selected consumer products and building materials. *Atmos. Environ.*, 37, 5485–5493 (2003)

99 Marklund, A., Andersson, B., Haglund, P.: Screening of organophosphorus compounds and their distribution in various indoor environments, *Chemosphere*, 53, 1137–1146 (2003)

100 Aston, L. S., Noda, J., Seibner, J. N., Reece, C. A.: Organophosphate flame-retardants in Needles of Pinus ponderosa in the Sierra Nevada foothills, *Bull. Environ. Contam. Toxicol.*, 57, 859–866 (1996)

101 Iuclid dataset for: Tris (2-chloro-1-methylethyl) phosphate, Akzo Nobel b.v., Amersfort, 15th March, 2001

102 Bester, K.: Tris (2-chloro-1-methylethyl) phosphate (TCPP) comparison of sludge and wastewater concentrations in German sewage plants, *J. Environ. Monitor.* 7, 509–513 (2005)

103 Sasaki, K., Suzuki, T., Takeda, M., Uchiyama, M.: Bioconcentration and excretion of phosphoric acid triesters by *killifish. Bull. Environ. Contam. Toxicol.*, *28*, 752–759 (1982)

104 Sandmeyer, E. E., Kirwin, C. J.: Esters. In: Clayton, G. D. & Clayton, F. E., ed. Patty's industrial hygiene and toxicology, 3rd revised ed., New York, Wiley-Interscience, Vol. 2A, 1981, pp. 2259–2412.

105 Windholz, ed.: The Merck index, 11th ed. Rahway, New Jersey, Merck and Co., Inc. (1983)

106 US EPA (Environmental Protection Agency, Office of Toxic Substances). Chemical hazard information profile draft report: tri(alkyl/ alkoxy) phosphates. Washington, DC, 1985.

107 Inchem. United nations environment program international labour organization, World health organization: International program on chemical safety, Environmental Health criteria 112, Tri-n-butylphosphate, Geneva, 1991

108 Inchem. United nations environment program international labour organization, World health organization: International program on chemical safety, Environmental Health criteria 209 Flame-retardants: Tris-(chloropropyl)phosphate and Tris-(2-chloroethyl)phosphate, Geneva, 1998

109 Huckins, J. N., Fairchild, J. F., Boyle, T. P.: Role of exposure mode in the bioavailability of triphenyl phosphate to aquatic organisms, *Arch. Environ. Contam. Toxicol.*, 21, 481–485 (1991)

110 Inchem. United Nations Environment Program International Labour Organization, World Health Organization: International Program on Chemical Safety, Environmental Health criteria 111, Triphenylphosphate, Geneva, 1991

111 Inchem. United Nations Environment Program International Labour Organization, World Health Organization: International Program on Chemical Safety, Environmental Health criteria 218, flame retartants (2003)

112 van Stee, L. L. P., Leonards, P. E. G., Vreuls, R. J. J., Brinkman, U. A. T.: Identification of non-target compounds using gas chromatography with simultaneous atomic emission and mass spectrometric detection, *Analyst*, 124, 1547–1552 (1999)

113 Kolpin, D. W., Furlong, E. T., Meyer, M. T., Thurman, E. M., Zaugg, S. D., Barber, L. B., Buxton, H. T.: Pharmaceuticals, hormones, and other organic wastewater contaminants in US streams, 1999–2000: A national reconnaissance, *Environ. Sci. Technol.*, 36, 1202–1211 (2002)

114 Fries, E., Puttmann, W.: Occurrence of organophosphate esters in surface water and ground water in Germany, *J. Environ. Monitor.*, 3, 621–626 (2001)

115 Prösch, J., Puchert, W., Gluschke, M.: Vorkommen von Chloralkylphosphaten in den Abläufen kommunaler Kläranlagen des deutschen Ostsee-Einzugsgebietes, *Vom Wasser*, 95, 87–96 (2000) (in German)

116 Kawagoshi, Y., Nakamura, S., Fukunaga, I.: Degradation of Organophosphoric esters in leachate from a sea based solid waste disposal site, *Chemosphere*, 48, 219–225 (2002)

117 van der Togt, B., van Ginkel, C. G.: Biodegradability of Fyrol PCF in the prolonged closed bottle test, Research Report (CRE EN F 02101), Akzo Nobel, Arnhem, December 2nd (2002)

118 Fischer, V.: Bestimmung von partikelgebundenen Schadstoffen mit besonderem Augenmerk auf Duftstoffen, phosphororganischen Flammschutzmitteln und Weichmachern im Abwasser mittels GC-MS, Diploma thesis Environmental Sciences, University Duisburg-Essen, 2006

119 Barcelo, D., Porte, C., Cid, J., Albaiges, J.: Determination of organophosphorus compounds in mediterranean costalwaters and biota samples using gas chromatography with nitrogen – phosphorus and chemical ionization mass spectrometric detection, *Int. J. Environ. Anal. Chem.*, 38, 199–209 (1990)

120 LeBel, G. L., Williams, D. T., Benoit, F. M.: Gaschromatographic determination of trialkyl/aryl phosphates in drinking water. following isolation using macroreticular resin, *J. Assoc. Off. Anal. Chem.*, 64, 991, 998 (1981)

121 Andresen, J. A., Grundmann, A., Bester, K.: Organophosphorus flame-retardants and plasticisers in surface waters, *Sci. Total Environ.*, 332, 1–3, 155–166 (2004)

122 Stackelberg, P. E., Furlong, E. T., Meyer, M. T., Zaugg, S. D., Henderson, A. K., Reissman, D B.: Persistence of pharmaceutical compounds and other organic wastewater contaminants in a conventional drinking-water-treatment plant, *Sci. Total Environ.*, 329, 99–113 (2004)

123 Heberer, T., Feldmann, D., Reddersen, K., Altmann, H. J., Thomas Zimmermann, T.: Production of drinking water from highly contaminated surface waters: removal of organic, inorganic, and microbial contaminants applying mobile membrane filtration units, *Acta Hydrochim. Hydrobiol.*, 30, 24–33 (2002)

124 Meyer, J. A., Bester, K.: Organophosphate flame-retardants and plasticisers in wastewater treatment plants. *J. Environ. Monitor.*, 6, 599–605 (2004)

125 Scott B. F, Sverko E, Maguire J.: Determination of benzothiazole and alkyl-phosphates in water samples from the Great Lakes drainage basin by gas chromatography/Atomic emission detection, *Water Qual. Res. J. Can.*, 31, 341–360 (1996)

126 Peck, A. M., Hornbuckle, K. C.: Synthetic Musk Fragrances in Lake Michigan, *Environ. Sci. Technol.*, 38, 367—372 (2004)

127 Berset, J. D., Bigler, P., Herren, D.: Analysis of nitro musk compounds and their amino metabolites in liquid sewage sludges using NMR and mass spectrometry, *Anal Chem.*, 72, 2124–2131 (2000)

128 Biselli, S., Gatermann, R., Kallenborn, R., Sydnes, L. K., Huehnerfuss, H.: Biotic and abiotic transformation pathways of synthetic musks in the aquatic environment. Handbook of Environmental Chemistry (3(Pt. X), 189–211. Springer, Berlin, Germany (2004)

129 Andresen, J., Bester, K.: Elimination of organophosphate ester flame-retardants and plasticizers in drinking water purification, *Wat Res.*, 40, 621–629 (2006)

130 Bester, K.: Fate of Triclosan and Triclosan-Methyl in sewage treatment plants and surface waters, Arch. *Environ. Contam. Toxicol.*, 49, 9–18 (2005)

131 ARGE Elbe (Ed.) Ausgewählte organische Spurenverunreinigungen in der Elbe und Elbeneben-flüssen im Zeitraum 1994–1999. Arbeitsgemeinschaft zur Reinhaltung der Elbe, Wassergütestelle Elbe, Hamburg (2002)

132 ARW (Ed.) Jahresbericht 1996–2001. Geschäftsstelle Arbeitsgemeinschaft Rhein-Wasserwerke, Köln (1997–2002)

133 Bundesamt für Seeschifffahrt und Hydrographie (BSH). Nordseezustand 2003. Bericht Nr. 38 (2005)

134 Glasmeyer, S. T., Furlong, E. T., Kolpin, D. W., Cahill, J. D., Zaugg, S. D., Werner, S. L., Meyer, M. T., Kryak, D. D.: Transport of Chemical and Microbial Compounds from Known Wastewater Discharges: Potential for Use as Indicators of Human Fecal Contamination, *Environ. Sci. Technol.*, 39, 5157–5169 (2005)

135 Wilson, B. A., Smith, V. H., DeNoyelles, F., Larive, C. K.: Effects of Three Pharmaceutical and Personal Care Products on Natural Freshwater Algal Assemblages, *Environ Sci Technol.*, 37, 1713–1719 (2003)

136 Valters, K., Li, H., Alaee, M., D'Sa, I., Marsh, G., Bergman, A., Letcher, R. L.: Polybrominated Diphenyl Ethers and Hydroxylated and Methoxylated Brominated and Chlorinated Analogues in the Plasma of Fish from the Detroit River, *Environ Sci Technol.*, 39, 5612–5619 (2005)

137 Anon. Ministerial declaration of the fourth international conference on the protection of the North Sea (1995), available on *http://odin.dep.no/md/nsc/declaration*

138 Föllmann, W., Spiteller, M., Degen, G. H., Vollmer, G., Bester, K.: Investigations of toxic effects of the flame-retardants Tris-(-2-chloroethyl)-phosphate (TCEP) and Tris-(2-chloro-propyl)-phosphate (TCPP) in vitro, *Naunyn-Schmiedeberg's Archives of Pharmacology*, 369, 578 (2004)

139 Gaido, K. W., Leonard, L. S., Lovell, S., Gould, J. C., Babai, D., Portier, C. J., McDonnell, D. P.: Evaluation of chemicals with endocrine modulating activity in a yeast-based steroid hormone receptor gene transcription assay, *Toxicology and Applied Pharmacology,* 143, 205–212 (1997)

140 European Commission, Joint Research Centre Institute for Health and Consumer Protection: Summary risk assessment report on 4,4'-isopropylidene-diphenol (Bisphenol A) (2004)

141 Environmental Protection Agency: National Drinking Water Regulations Technical Fact Sheet on Di (2-Ethylhexyl) Phthalate, January (1998)

142 Bester K., Andresen J., Schlüsener M. P.: Bestimmung von Abbaubilanzen von Substanzen aus personal-care Produkten und Hormonen in Kläranlagen (BASPiK), final report (2005).

143 Bund/Länderausschuß für Chemikaliensicherheit (BLAC); Arzneimittel in der Umwelt, Auswertung der Untersuchungsergebnisse, BLAG-AG, Hamburg (2003)

144 Ministry for the Environment and Conservations, Agriculture and Consumer Protection of the state of North Rhine Westphalia; Untersuchungen zum Eintrag und zur Elimination von gefährlichen Stoffen in kommunalen Kläranlanlagen, final report (2004).

145 Ternes, T. A., Kreckel, P., Mueller, J.: Behaviour and occurrence of estrogens in municipal sewage treatment plants – II. Aerobic batch experiments with activated sludge, *Sci. Tot. Environ.,* 225, 91–99 (1999)

146 Thiele, B., Günther, K., Schwuger. M. J.:. Alkyphenolethoxylates: Trace Analysis and Environmental Behaviour, *Chem. Rev.,* 97, 3247–3272 (1997)

147 Lee P. C., Lee, W.: In vivo estrogenic action of nonylphenol in inmature female rats, *Bull. Environ. Contam. Toxicol.,* 7, 341–348 (1996)

148 Fields, J. A., Reed, R. L.: Nonylphenol polyethoxycarboxylate metabolites of nonionic surfactants in US paper mill effluents, munincipal sewage treatment plant effluents, and river water, *Environ. Sci. Technol.,* 30, 3544–3550 (1996)

149 Ahel, M., Scully jr., F. E., Hoigne, J., Giger, W.: Photochemical degradation of nonylphenol and nonylphenol ethoxylates in Natural Waters, *Chemosphere.* 28, 1361–1368 (1994)

150 Maki, H., Fujita M., Fujiwara, Y.: Identification of final biodegradation product of nonylphenolethoxylate (NPE) by river microbial consortia, *Bull. Environ. Contam. Toxicol.,* 57, 881–887 (1996)

151 Comber, M. H. J., Williams T. D., Steward, K. M.: The effects of nonylphenol on Daphnia Magna, *Water Res.,* 27, 273–276 (1993)

152 Bennie, D. T., Sullivan, C. A., Lee, H. B., Peart, T. E., Maguire, R. J.: Occurance of alkylphenols and alkylphenolmono- and diethoxylates in natural waters of the laurentian Great Lakes basin and the upper St. Lawrence River, *Sci. Total Environ.,* 193, 263–275 (1997)

153 Theobald, N., Lange, W., Gählert, W., Remmer, F.: Mass spectrometric investigations of water extracts of the river Elbe for the determination of potential inputs of pollutants into the North Sea, *Fresenius J. Anal. Chem.*, 353, 50–56 (1995)

154 Kvestak, R., Ahel, M.: Biotransformation of nonylphenolpolyethoxylate surfactants by estuarine mixed bacterial cultures, *Arch. Environ. Contam. Toxicol.*, 29, 551–556 (1995)

155 Kvestac, R., Terzic S., Ahel, M.: Input and distribution of alkylphenol poly-ethoxylates in a stratified estuary, *Mar. Chem.*, 46, 89–100 (1994)

156 Marcomini, A., Pavoni, B., Sfriso A., Orio, A. A.: Persistent metabolites of alkylphenol polyethoxylates in the marine environment, *Mar. Chem.*, 29, 307–323 (1990)

157 Valls, M., Bayona, J. M., Albaiges, J.: Broad spectrum analysis of ionic and non-ionic organic contaminants in urban wastewaters and coastal receiving aquatic systems, *Int. J. Environ. Anal. Chem.*, 39, 329–348 (1990)

158 Marcomini, A., Pojana, G., Sfriso, A., Alonso, J.-M. Q.: Behavior of anionic and nonionic surfactants and their persistent metabolites in the Venice Lagoon, Italy, *Environ. Toxicol. Chem.*, 19, 2000–2007 (2000)

159 Kannan, N., Yamashita, N., Petrick, G., Duinker, J. C.: Polychlorinated Biphenyls and nonylphenols in the Sea of Japan, *Environ. Sci. Technol.*, 32, 1747–1753 (1998)

160 Müller, S., Efer, J., Engewald, W.: Water pollution screening by large-volume injection of aqueous samples and application to GC/MS [gas chromatog./ mass spectrometry] analysis of a river Elbe sample, *Fresenius J. Anal. Chem.*, 357, 558–560 (1997)

161 Guzzella, L., Sora, S.: Mutagenic activity of lake water samples used as drinking water resources in Northern Italy, *Water Res.*, 32, 1733–1742 (1998)

162 Olsson, A., Bergman, A.: A new persistent contaminant detected in Baltic Wildlife: Bis (4-chlorophenyl)-sulfone, *Ambio*, 24, 119–123 (1995)

163 Bester, K., Lange, W., Theobald, N.: Results of non target screening of lipo-philic organic pollutants in the German Bight V: Xanthen-9-on, *Water Res.*, 34, 2277–2282 (2000)

164 Bester, K., Biselli, S., Gatermann, R., Hühnerfuss, H., Lange, W., Theobald, N.: Results of non target screening of lipophilic organic pollutants in the German Bight III: 2,5-Dichloroaniline, *Chemosphere*, 36, 1973–1983 (1998)

165 Theobald, N., Gaul, H., Ziebarth, U.: Verteilung von organischen Schadstof-fen in der Nordsee und angrenzenden Seegebieten. (Distribution of organic pollutants in the North Sea and adjoining areas), *Dt. Hydrogr. Z., Suppl.*, 6, 81–93 (1996)

166 Bester, K., Gatermann, R., Hühnerfuss, H., Lange, W., Theobald, N.: Re-sults of non target screening of lipophilic organic pollutants in the German Bight IV: Identification and Quantification of Chloronitrobenzenes and Dichloronitrobenzenes, *Environ. Pollut.*, 102, 163–169 (1998)

167 Franke, S., Hildebrandt, S., Francke, W., Bester, K., Hühnerfuss, H., Gater-mann, R.: Chlorinated Bis(propyl)ethers and chlorinated Bis(ethyl)formals in the German Bight of the North Sea, *Mar. Pollut. Bull.*, 36, 546–551 (1998)

168 Bester, K., Theobald, N., Schröder H.Fr.: Nonylphenols, Nonylphenol-ethoxylates, Linear Alkylbenzenesulfonates (LAS) and Bis (4-chlorophenyl)-sulfone in the German Bight of the North Sea, *Chemosphere*, 45, 817–826 (2001)

169 Ekelund, R., Granmo, A., Magnusson, K., Berggren, M., Bergman, A.: Bio-degradation of 4-nonylphenol in seawater and sediment, *Environ. Pollut.*, 79, 59–61 (1993)

170 Koopman, G., Voppel, D., Rühl, N.P., Heinrich, H.: Transport, Umsatz und Veriabilität von Schadstoffen in der Deutschen Bucht 1990–1992, BSH-BMFT Report 03F0559A (in German) (1994)

171 Nice, H.E., Thorndyke, M.C., Morritt, D., Steele, S., Crane, M.: Development of Crassostrea gigas larvae is affected by 4-nonylphenol, *Mar. Pollut. Bull.*, 40, 491–496 (2000)

172 Billinghurst, Z., Clare, A.S., Fileman, T., Mcevoy, M., Readman, J., Depledge, M.H.: Inhibition of Barnacle settlement by the environmental oestrogen 4-nonylphenol and the natural oestrogen 17β-oestradiol, *Mar. Pollut. Bull.*, 36, 833–839 (1998)

173 Billinghurst, Z., Clare, A.S., Matsumura, K., Depledge, M.H.: Induction of cypris major protein in barnacle larvae by exposure to 4-n-nonylphenol and 17β-oestradiol, *Aquat. Toxicol.*, 47, 203–212 (2000)

174 Schröder, H.Fr., Ventura, F.: Applications of liquid chromatography-mass spectrometry in environmental chemistry; Characterization and determination of surfactants and their metabolites in water samples by modern mass spectrometric techniques, in: Techniques and Instrumentation in Analytical Chemistry – Vol. 21; Sample Handling and Trace Analysis of Pollutants – Techniques, Applications and Quality Assurance. D. Barceló, (Ed.), pp. 828–933. Elsevier, Amsterdam (2000)

175 Schröder, H.Fr.: Analysis of polar organic pollutants in the Elbe river by flow injection analysis and high-performance liquid chromatography with tandem mass spectrometry, *J. Chromatogr. A*, 777, 127–139 (1997)

176 Schröder, H.Fr.: Biochemisch schwer abbaubare organische Stoffe in Ab-wässern und Oberflächenwässern – Vorkommen, Bedeutung und Elimina-tion, *Gewässerschutz, Wasser, Abwasser, GWA*, 166, Herausg.: Dohmann, Aachen, ISBN 3-932590-43-0. Habilitationsschrift vorgelegt im Fachbereich 3, Fakultät für Bauingenieur- und Vermessungswesen der RWTH Aachen (1997)

177 González-Mazo, E., Honig, M., Barceló, D., Gómez-Parra, A.: Monitoring long-chain intermediate products from the degradation of linear alkylben-zene sulfonates in marine environment by solid-phase extraction followed by liquid chromatography/ionspray mass spectrometry, *Environ. Sci. Tech-nol.*, 31, 504–510 (1997)

178 Olsson, A., Bergman, A.: A new persistent contaminant detected in Baltic Wildlife: Bis (4-chlorophenyl)-sulfone, Ambio, 24, 119–123 (1995)

179 Achte Verordnung zur Änderung chemikalienrechtlicher Verordnungen, *BGBL*, I, 9, 328–330 (2004)

180 European Parliament: Directive 2003/53/EC of the European Parliament and of the Council of 18 June 2003 amending for the 26th time Council Directive 76/769/EEC relating to restrictions on the marketing and use of certain dangerous substances and preparations (nonylphenol, nonylphenol ethoxylate and cement), *Official Journal of the European Union*, L 178, 24–27 (2003)

181 Spies, R. B., Andresen, B. D., Rice, D. W. Jr.: Benzothiazoles in estuarine sediments as indicator of street runoff, *Nature*, 327, 697–699 (1987)

182 Janssen, D.: Gas chromatographic-mass spectrometric investigations on lipophilic anthropogenic substances in environmental samples II A splitter for the simultaneous detection of capillary gas chromatograms by up to four different GC-Detectors, *Fresenius Z. Anal. Chem.*, 331, 20–26 (1988)

183 Bester, K., Hühnerfuss, H., Brockmann U., Rick H. J.: Biological effects of triazine herbicide contamination on marine phytoplankton, *Arch. Environ. Contam. Toxicol.*, 29 277–283 (1995)

184 Hendriks, A. J., Maas-Diepveen, J. L., Noordsig A., Van der Gaag, M. A.: Monitoring response of XAD-concentrated water in the Rhine delta, A major part of the toxic compounds remains unidentified, *Water Res.*, 28, 581–598 (1994)

185 Parberg, C., Taylor C. D.: Determination of methylene(bisthiocyanate) and 2-thiocyanatomethylthiobezo(d)thiazol in leather process liquors by high-performance liquid chromatography, *Analyst*, 114, 361–363 (1989)

186 Daniels, C. R., Swan E. P.: HPLC assay of the anti-stain chemical TCMBT applied to lumber surfaces, *J. Chromatogr. Sci.*, 25, 43–45 (1987)

187 Health and safety executive, London, Pesticides 1990, London, 1990

188 Reemtsma, T., Fiehn, O., Kalnowski, G., Jekel, M.: Microbial transformations and biological effects of fungicide derived benzothiazoles determined in industrial wastewater, *Environ. Sci. Technol.*, 29, 478–485 (1995)

189 Perraud, R., Papazian M., Krahe E.: Zersetzung biozider Verbindungen in wäßrigem Millieu, *GIT*, 4, 312–316 (1995)

190 Metcalfe, J. L., Fox M. E., Carey J. H.: Freshwater leeches (hirudinea) as a screening tool for detecting organic contaminations in the environment, *Environ. Monitor. Assess.*, 11, 147–169 (1988)

191 Welch, D. T., Watts, C. D.: Collection and identification of trace organic compounds in atmospheric deposition from a semi rural site in the UK, Intern. J. Anal. Chem., 38 185–198 (1990)

192 Runge G., Steinhart, H.: Determination of volatile sulfur compounds in the edible part of carp, *Agribiol. Res.*, 43, 155–163 (1990)

193 Sauer J., Antusch E., Ripp C.: Monitoring of lipophilic organic pollutants in sewer systems by sewer slime analysis, *Vom Wasser*, 88, 49–69 (1997)

194 Fromme H., Otto T., Pilz K., Neugebauer, F.: Levels of synthetic musks; Bromocyclene and PCBs in eel (*Anguilla anguilla*) and PCBs in sediment samples from some waters of Berlin, Germany, *Chemosphere*, 39, 1723–1735 (1999)

195 Ahel, M., Giger, W.: Determination of alkylphenols and alkylphenol mono- and diethoxylates in environmental samples by high-performance liquid chromatography, *Anal.Chem.*, 57, 1577–1588 (1985)

196 Aherne, G. W., Briggs, R.: The relevance of the presence of certain synthetic steroids in the aquatic environment, *J. Pharm. Pharmacol.*, 41, 735–736 (1989)

197 Goodman Gilman, A., Rall, T. W., Nies, A. S., Taylor, P.: Goodman and Gillman's in The pharmacological basis of therapeutics, Pergamon Press, New York, United States (1990)

198 Sumpter, J. P.: Xenoendocrine disrupters – environmental impacts, *Toxicol. Lett.*, 102–103, 337–342 (1998)

199 Robinson, C. D., Brown, E., Craft, J. A., Davies, I. M., Moffat, C. F., Pirie, D., Robertson, F., Stagg, R. M., Struthers, S.: Effects of sewage effluent and ethynyl oestradiol upon molecular markers of oestrogenic exposure, maturation and reproductive success in the sand goby (*Pomatoschistus minutus*, Pallas), *Aquat. Toxicol.*, 62, 119–134 (2003)

200 Routledge, E. J., Sheahan, D., Desbrow, C., Brighty, G. C., Waldock, M., Sumpter, J. P.: Identification of estrogenic chemicals in STW effluent. 2. In Vivo Responses in Trout and Roach, *Environ. Sci. Technol.*, 32, 1559–1565 (1998)

201 Neu, H. C.: The crises in antibiotics resistance, *Science*, 257, 1064–1073 (1992)

202 Ternes, T. A., Stumpf, M., Mueller, J., Haberer, K., Wilken, R. D., Servos, M.: Behavior and occurrence of estrogens in municipal sewage treatment plants – I. Investigations in Germany, Canada and Brazil, *Sci. Tot. Environ.*, 225, 81–99 (1999)

203 Holger, M., Kuch, K. B.: Determination of endogenous and exogenous estrogens in effluents from sewage treatment plants at the ng L–1-level, *Fresenius J. Anal. Chem.*, 366, 392–395 (2000)

204 Ternes, T. A., Andersen, H., Gilberg, D., Bonerz, M.: Determination of estrogens in sludge and sediments by liquid extraction and GC/MS/MS, *Anal. Chem.*, 74, 3498–3504 (2002) P

205 Belfroid, A. C., Van der Horst, A., Vethaak, A. D., Schäfer, A. J., Rijs, G. B., Wegener, J., Cofino, W.P.: Analysis and occurrence of estrogenic hormones and their glucuronides in surface water and wastewater in The Netherlands, *Sci. Tot. Environ.*, 225, 101–108 (1999)

206 Lagana, A., Bacaloni, A., Fago, G., Marino, A.: Trace analysis of estrogenic chemicals in sewage effluent using liquid chromatography combined with tandem mass spectrometry, *Rapid. Commun. Mass Spectrom.*, 14, 401–407 (2000)

207 Zühlke, S., Duennbier, U., Heberer, T.: Determination of estrogenic steroids in surface water and wastewater by liquid chromatography – electrospray tandem mass spectrometry, *J. Sep. Sci.*, 28, 52–58 (2005)

208 Göbel, A., McArdell, C. S., Suter, M. J.-F., Giger, W.: Trace determination of macrolide and sulfonamide antimicrobials a human sulfonamide metabolite and trimetoprim in wastewater using liquid chromatography coupled to electrospray tandem mass spectrometry, *Anal. Chem.*, 76, 4756–4764 (2004)

209 Hirsch, R., Ternes, T., Haberer, K., Kratz, K.-L.: Occurrence of antibiotics in the aquatic environment, *Sci. Tot. Environ.*, 225, 109–118 (1999)

210 European Commission: Implementing Council Directive 96/23/EC concerning the performance of analytical methods and the interpretation of results, *Official J.*, L. 221, 8–36 (2002)

211 Schlüsener, M. P., Bester, K., Spiteller, M.: Determination of antibiotics such as macrolides, ionophores and tiamulin in liquid manure by HPLC–MS/MS, *Anal. Bioanal. Chem.*, 375, 942–947 (2003)

212 Yee-Chung, M., Hee-Yong, K.: Determination of steroids by liquid chromatography/mass spectrometry, *J. Am. Soc. Mass Spectrom.*, 8, 1010–1020 (1997)

213 Bartonti, C., Curini, R., D'Ascenzo, G., Di Corcia, A., Gentili, A., Saperi, R.: Monitoring natural and synthetic estrogens at activated sludge sewage treatment plants and in a recieving river water, *Environ. Sci. Technol.*, 34, 5059–5066 (2000)

214 Gentili, A., Perret, D., Marchese, S., Mastropasqua, R., Curini, R., Di Corisa, A.: Analysis of free estrogens and their conjugates in sewage and river waters by solid-phase extraction then liquid chromatography-electrospray-tandem mass spectrometry, *Chromatographia*, 56, 25–32 (2002)

215 Pfeifer, T., Tuerk, J., Bester, K., Spiteller, M.: Determination of selected sulfonamide antibiotics and trimethoprim in manure by electrospray and atmospheric pressure chemical ionization tandem mass spectrometry, *Rapid Commun. Mass Spectrom.*, 16, 663–669 (2002)

216 Stüber, M., Reemtsma, T.: Evaluation of three calibration methods to compensate matrix effects in environmental analysis with LC-ESI-MS, *Anal. Bioanal. Chem.*, 378, 910–916 (2004)

217 Rodil, R., Quintana, J. B., Reemtsma, T.: Liquid chromatography-tandem mass spectrometry determination of nonionic organophosphorus flame-retardants and plasticizers in wastewater samples, *Anal. Chem.*, 77, 3083–3089 (2005)

218 Isobe, T., Shiraishi, H., Yasuda, M., Shinoda, A., Suzuki, H, Morita, M.: Determination of estrogens and their conjugates in water using solid-phase extraction followed by liquid chromatography-tandem mass spectrometry, *J. Chromatogr. A.*, 984, 195–202 (2003)

219 Gaul, H., Ziebarth, U.: Methods for the analysis of lipophilic components in water and results about the distribution of different organochlorine compounds in the North Sea, *Dt. Hydrogr. Z.*, 36, 191–212 (1983)

220 Theobald, N., Lange, W., Rave, A., Pohle U., Koennecke, P.: Ein 100-l-Glaskugelschöpfer zur kontaminationsfreien Entnahme von Seewasser für die Analyse lipophiler organischer Stoffe, *Dt. Hydrogr. Z.*, 43, 311–322 (1990)

221 Di Corcia, A., Marchetti, M.: Multiresidue method for pesticides in drinking water using a graphitized carbon Black Cartrige, *Anal. Chem.*, 63, 580–585 (1991)

222 Dav, L.M., Baldi, M., Penazzi, L., Liboni, M.: Evaluation of the membrane approach to solid phase extractions of pesticide residues in drinking water, *Pestic. Sci.*, 35, 63–67 (1992)

223 Beltran, J., López, F.J., Hernández, F.: Solid-Phase Extraction of Pesticide-Residues from Ground Water – Comparison Between Extraction Cartridges and extraction disks, *Anal. Chim. Acta*, 283, 297–303 (1993)

224 Brooks, M.W., Tessier, D. Soderstrom, D., Jenkins, J., Clark, J.M.: A rapid method for the simultaneous analysis for chlorpyrifos, isofenphos, carbaryl, iprodione, and triadimefon in groundwater by solid-phase extraction. *J. Chromatogr. Sci.*, 28, 487–489 (1990)

225 Fiehn, O., Jekel, M.: Comparison of sorbents using semipolar to highly hydrophilic compounds for a sequential solid-phase extraction procedure of industrial wastewaters *Anal. Chem.*, 68, 3083–3089 (1996)

226 Lacorte, S., Barceló, D.: Improvements in the determination of organophosphorus pesticides in ground- and wastewater samples from interlaboratory studies by automated on-line liquid-solid extraction followed by liquid chromatography diode array detection, *J. Chromatogr. A*, 725, 85–92 (1996)

227 Louter, A.J.H., van Beekvelt, C.A., Cid Montanes, P., Slobodnik, J., Vreuls, J.J., Brinkman, U.A.Th.: Analysis of microcontaminants in aqueous samples by fully automated on-line solid-phase extraction gas chromatography mass selective detection, *J. Chromatogr. A*, 725, 67–83 (1996)

228 Fernández, M., Ibáñez, M., Picó, Y., Mañes, J.: Spatial and temporal trends of paraquat, diquat, and difenzoquat contamination in water from marsh areas of the Valencian community (Spain), *Arch. Environ. Contam. Toxicol.*, 35, 377–384 (1998)

229 Aguilar, C., Ferrer, I., Borrul, F., Marce, R.M., Barceló, D.: Monitoring of pesticides in river water based on samples previously stored in polymeric cartridges followed by on-line solid-phase extraction liquid chromatography diode array detection and confirmation by atmospheric pressure chemical ionization mass spectrometry, *Anal. Chim. Acta*, 386, 237–248 (1999)

230 Chiron, S., Martinez, E., Barceló, D.: On-line and offline sample preparation of acidic herbicides and bentazone transformation products in estuarine waters, *J. Chromatogr. A*, 665, 283–293 (1994)

231 Blackburn, M.A., Kirby, S.J., Waldock, M.J.: Concentrations of alkyphenol polyethoxylates entering UK estuaries, *Mar. Pollut. Bull.*, 38, 109–118 (1999)

232 Zhou, J.L, Fileman, T.W., House, W.A., Long, J.L.A., Mantoura, R.F.C., Meharg, A.A., Osborne, D., Wright, J.: Fluxes of organic contaminants from the river catchment into, through and out of the Humber estuary, UK, *Mar. Pollut. Bull.*, 37, 330–342 (1998)

233 Utvik, T.I.R., Durell, G.S., Johnsen, S.: Determining produced water originating polycyclic aromatic hydrocarbons in North Sea waters: Comparison of sampling techniques, Mar. Pollut. Bull., 38, 977–989 (1999)

234 Galceran, M.T., Jauregui, O.: Determination of phenols in sea water by liquid-chromatography with electrochemical detection after enrichment by using solid phase extraction cartridges and disks, *Anal. Chim. Acta,* 304, 75–84 (1995)

235 Moltó, J.C., Albelda, C., Font, G., Mañes, J.: Solid-phase extraction of organochlorine pesticides from water samples, *Intern. J. Environ. Anal. Chem.,* 41, 21–26 (1990)

236 Tolosa, I., Readman, J.W., Mee, L.D.: Comparison of the performance of solid phase extraction techniques in recovering organophosphorus and organochlorine compounds from water, *J. Chromatogr. A,* 725, 93–106 (1996)

237 Lacorte, S., Molina, C., Barceló, D.: Screening of organophosphorus pesticides in environmental matrices by various gas-chromatographic techniques, *Anal. Chim. Acta,* 281, 71–84 (1993)

238 Schmidt, L., Sun, J.J., Fritz, J.S., Hagen, D.F., Markell, C.G., Wisted, E.E.: Solid phase extraction of phenols using membranes loaded with modified polymeric resins, *J. Chromatogr.,* 641, 57–61 (1993)

239 Renner, T., Baumgarten, D., Unger K.K.: Analysis of organic pollutants in water at trace levels using fully automated solid-phase extraction coupled to high-performance liquid chromatography, *Chromatographia,* 45, 199–205 (1997)

240 Hagen, D.F., Markell, C.G., Schmitt, G.A., Blevins, D.D.: Membrane approach to solid-phase extractions, *Anal. Chim. Acta.,* 236, 157–164 (1990)

241 Alonso, M.C. Barceló, D.: Tracing polar benzene- and naphthalenesulfonates in untreated industrial effluents and water treatment works by ion-pair chromatography-fluorescence and electrospray-mass spectrometry, *Anal. Chim. Acta,* 400, 211–231 (1999)

242 Loos, R., Niessner, R.: Analysis of aromatic sulfonates in water by solid-phase extraction and capillary electrophoresis, *J. Chromatogr. A,* 822, 291–303 (1998)

243 Kira, S., Sakano, M., Nogami, Y.: Measurement of a time-weighted average concentration of polycyclic aromatic hydrocarbons in aquatic environment using solid phase extraction cartridges and a portable pump, *Bull. Environ. Contam. Toxicol.,* 58, 878–884 (1997)

244 Di Corcia, A., Samperi, R., Marcomini, A., Stelluto, S.: Graphitized carbon black extraction cartridges for monitoring polar pesticides in water, *Anal. Chem.,* 65, 907-912 (1993)

245 Böhme, C., Schmidt, T.C., von Löw, E., Stork, G.: Solid-phase extraction of aminobenzoic acids and aminotoluenesulfonic acids with graphitized carbon black, *Fresenius J. Anal. Chem.,* 360, 805–807 (1998)

246 Font, G., Mañes, J., Moltó, J.C., Picó, Y.: Solid-phase extraction in multiresidue pesticide analysis of water, *J. Chromatogr.,* 642, 135–161 (1993)

247 Benfenati, E., Tremolada, P., Chiappetta, L., Frassanito, R., Bassi, G., Di Toro, N., Fanelli, R., Stella, G.: Simultaneous analysis of 50 pesticides in water samples by solid phase extraction and GC – MS, *Chemosphere,* 21, 1411–1421 (1990)

248 Junk, G. A., Richard, J. J.: Organics in water: Solid Phase extraction on a small scale, *Anal. Chem.*, 60, 451–454 (1988)

249 Tolosa, I., Readman, J. W., Mee, L. D.: Comparison of the performance of solid phase extraction techniques in recovering organophosphorus and organochlorine compounds fron water, *J. Chromatogr. A*, 725, 93–106 (1996)

250 Hinckley, D. A., Bidleman, T. F.: Analysis of pesticides in seawater after enrichment onto C8 bonded-phase cartriges, *Environ. Sci. Technol.*, 23, 995–1000 (1989)

251 Ternes, T. A., Stumpf, M., Schuppert, B., Haberer, K.: Simultaneous determination of antiseptics and acidic drugs in sewage and river water, *Vom Wasser*, 90, 295–309 (1998)

252 Barceló, D., Chiron, S., Lacorte, S., Martinez, E., Salau, J. S., Hennion, M. C.: Solid phase sample preparation and stability of pesticides in water using empore disks, *Trends in Analytical Chemistry*, 13, 352–361 (1994)

253 Renner, T., Baumgarten, D., Unger, K. K.: Analysis of organic pollutants in water at trace levels using fully automated solid-phase extraction coupled to high-performance liquid chromatography, *Chromatographia*, 45, 199–205 (1997)

254 Pichon, V., Coumes, C. C. D., Chen, L., Guenu, S., Hennion, M. C.: Simple removal of humic and fulvic acid interferences using polymeric sorbents for the simultaneous solid-phase extraction of polar acidic, neutral and basic pesticides, *J. Chromatogr. A*, 737, 25–33 (1996)

255 Schilling, R., Clarkson, P. J., Cooke, M.: Enhanced recovery of chlorophenols from surface waters using polymer based extraction cartridges, *Fresenius J. Anal. Chem.*, 360, 90–94 (1998)

256 Schulz-Bull, D. E., Petrick, G., Kannan, N., Duinker, J. C.: Distribution of individual chlorobiphenyls (PCB) in solution and suspension in the Baltic Sea, *Mar. Chem.*, 48, 245–270 (1995)

257 Gómez-Belinchón, J. I., Grimalt, J. O., Albaigés, J.: Intercomparison study of liquid-liquid extraction on polyurethane and amberlite XAD-2 for the analysis of hydrocarbons, polychlorobiphenyl, *Environ. Sci. Technol.*, 22, 677–685 (1988)

258 Ehrhardt, M. G., Burns, K. A.: Petroleum-derived dissolved organic compounds concentrated from inshore waters in Bermuda, *J. Exp. Mar. Biol. Ecol.*, 138, 35–47 (1990)

259 Schulz-Bull, D. E., Petrick, G., Duinker, J. C.: Polychlorinated biphenyls in North Sea water, *Mar. Chem.*, 36, 365–384 (1991)

260 Petrick, G., Schulz-Bull, D. E., Martens, V., Scholz, K., Duinker, J. C.: An in-situ filtration/extraction system for the recovery of trace organics in solution and on particles tested in deep ocean water, *Mar. Chem.*, 54, 97–105 (1996)

261 Sturm, B., Knauth, H.-D., Theobald, N., Wünsch, G.: Hydrophobic organic micropollutants in samples of coastal waters: efficiencies of solid-phase extraction in the presence of humic substances, *Fresenius J. Anal. Chem.*, 361, 803–810 (1998)

262 Franke, S., Hildebrandt, S., Schwarzbauer, J., Link, M., Francke, W.: Organic compounds as contaminants of the Elbe river and its tributaries part II, GC-MS screening for contaminants of the Elbe water, *Fresenius J. Anal. Chem.*, 353, 39–49 (1995)

263 Specht, W., Tilkes, M.: Gas-chromatographische Bestimmung von Rückständen an Pflanzenbehandlungsmitteln nach clean up über Gelchromatographie und Mini-Kieselgel Säulen-Chromatographie, *Fresenius Z. Anal. Chem.*, 322, 443–455 (1985)

264 Noble, A.: Partition coefficiants (n-octanol water) for pesticides, *J. Chromatogr.*, 642, 3–14(1993)

265 Verschueren, K.: Handbook of Environmental Data on Organic Chemicals, ITP, New York, 1996

266 Bester, K., Hühnerfuss, H.: Improvements of a combined size exclusion chromatography and solid phase extraction approach for the clean up of marine sediment samples for trace analysis of pesticides, *Fresenius J. Anal. Chem.*, 358, 630–634 (1997)

267 Führer, U., Deißler, A., Ballschmiter, K.: Determination of biogenic halogenated methyl-phenyl ethers (halogenated anisoles) in the picogram m-3 range in the air, *Fresenius J. Anal. Chem.*, 354, 333-343 (1996)

268 Oehmichen, U., Haberer, K.: Stickstoffherbizide im Rhein, *Vom Wasser*, 66, 225–241 (1986)

269 Bester, K., Hühnerfuss, H., Brockmann, U., Rick, H.J.: Biological effects of triazine herbicide contamination on marine phytoplancton, *Arch. Contamin. Toxicol.*, 23, 277–283 (1995)

270 Weigel, S., Bester, K., Hühnerfuss, H.: A new method for rapid solid-phase extraction of large volume water samples and its application in the screening of North Sea water for organic contaminants by GC/MS, *J. Chromatogr.*, A 912, 151–161 (2001)

271 Hühnerfuss, H., Bester, K., Landgraff, O., Pohlmann, T., Selke, K.: Annual balances of hexachlorocyclohexanes, polychlorinated biphenyls, and triazines in the German Bight, *Mar. Pollut. Bull.*, 34, 419–426 (1997)

272 Weigel, S., Bester, K., Hühnerfuss, H.: Identification and quantification of polar pesticides, industrial chemicals, and organobromine compounds in the North Sea: dichlobenil, dichloropyridine, bromoindols and others, *Marine Pollut Bull.*, 50, 252–263 (2005)

273 Bester, K., Hühnerfuss, H., Lange, W., Rimkus, G.G., Theobald, N.: Results of non-target screening of lipophilic organic pollutants in the German Bight II: Polycyclic Musk Fragrances, *Water Research*, 32, 1857–1863 (1998)

Subject Index

a
activated carbon filter 103
activated carbon filtration 107
aeration basin 22, 137, 140
AHTN 9
4-amino musk xylene 126
anhydro-erythromycin 180
annual transport 100
APCI 156
aspirin IX

b
balance 58
bank filtration 103
bed height 201
benzophenone 3 69
benzothiazoles 164 ff.
biodegradation 1
biological treatment 19
biotransformation 17, 36
bis(4-chlorophenyl)-sulfone 158
bisphenol A 134
bound residues 69
bromocyclene 172 ff.

c
caffein 198
carbamazepine 213
chlorinated ethers 213
chloropyridine 213
clarithromycin 136
cyclopentadecanolide 51

d
degradation 127
DEHP 91
denitrification 3, 139
2,4-dibromoanisol 198
dibutyl phthalate 134
dichloropyridines 198
diethylstilbestrol (DES) 129, 133
dilution 127
dilution factor 120
discharge 209
diurnal cycle 147
drinking water 101 ff.
dry weather 143

e
effluent 16, 81, 192
Elbe estuary 119
Elbe river 38
electron capture detection 5
elimination constants 67
elimination rate constant 99
elimination rates 88, 128
emission rates 69
Emscher river 31
enantiomeric ratios 36
enantioselective analysis 1
enantioselective HPLC 7
enantioselective determination
    1172
enantioselective gas
    chromatography 4
enantioselectivity 4
endocrine properties 70 ff.

Personal Care Compounds in the Environment: Pathways, Fate, and Methods for Determination. Kai Bester
Copyright © 2007 WILEY-VCH Verlag GmbH & Co. KGaA, Weinheim
ISBN: 978-3-527-31567-3

endocrine-disrupting compounds 128
erythromycin 135, 144
E-SCREEN 70
estradiol 145
β-estradiol 3-sulfate 132, 142
17-β-estradiol 129, 131
estriol 129, 131
estrone 131, 145
estrone-3-sulfate 132
17-α-ethinyl estradiol 132
ethylenbrassylate 51
ethylhexyl methoxycinnamate 69

f
false positive 187
filter candles 201
final filter 61
final sedimentation tank 19
fish 175
flame-retardants 74

g
German Bight 161, 213
Glatt river 43
granular activated carbon 111
grit chamber 139

h
habanolide 51
half-life 99, 219
Hamilton Harbor 116, 126
HCH 171
HHCB 9, 122, 206
HHCB-lactone 9, 32, 122
HPLC-MS/MS 179 ff.
human milk 54
humic compounds 78
hydrodex 5
hydroxyestrone 141, 146

i
immission scenarios 37
influent 16, 81, 192
input to STP 209

intermediate settling tank 19
internal standard calibration 185
isotope dilution 188

l
Lake Baldeney 33
Lake Kemnaden 33, 96
Lake Kettwig 33
Lake Ontario 116, 124
large-volume injection 104
limit of quantification 12
linear alkylbenzene sulfonates (LAS) 154
lipodex C 5
lipodex E 5
lipophilic film 28
Lippe river 31
liquid–liquid extraction 95
LOQ 105
lyophilized sludge 205

m
macrolide antibiotics 147
management options 208
marinas 160
marine ecosystem 216
marine environment 113
marine sediments 161
marine water samples 194
mass balance assessment 13 ff., 79 ff.
mass spectrometry 5
matrix calibration 188
matrix effects 187 ff.
MCF cells 70
mecoprop 6
mestranol 133
methyl triclosan 57, 126, 220
methylation processes 61
4-methylbenzylidene-camphor 69
methylthiobenzothiazole 43, 164
metolachlor 202
mineralization 1
multilayer filter 112
multiple reaction monitorings 180
multi-residue method 180

musk ketone   44, 46, 51
musk xylene   44, 46, 51
musk xylene-4-amine   46

n

N,N-diethyl-3-tolunamid (DEET)   198
neutralization line   138
nitroaromatic musk fragrances   124
4-nonylphenyl   129, 153 ff.
nonylphenol polyethoxylates   154
North Sea   38 ff., 116, 119 ff.
NPEO   154

o

Oslo and Paris Commissions   154
OSPRCOM   2
OTNE   44 ff., 217
Our Stolen Future   IX
oxidation procedures   210
oxidative biotransformation   208
oxidative transformation   3
ozonization   111

p

particulate transports   91 ff.
per-capita emissions   50
personal care compounds   122
pesticides   4
phase transfer of pollutants   6
phosphate precipitation   137, 140
plastic cutting boards   54
plasticizers   74
polycyclic musk compounds   9
polymeric material   91
polyurethane foam plates   80
POPs   211
poststream denitrification   136
precipitation   3
preliminary settling tank   136
primary settling tank   138
primicarb   202
proton exchange   28

q

quality assurance data   105

r

rain events   143
rainfalls   143
recovery   191
recovery rates   12
reductive transformation   3, 210
relative standard deviation   191
remobilization   127
Rhine river   19, 37
roxithromycin   135
RSD   105
Ruhr megapolis   10
Ruhr river   25, 37

s

sample transport   190
sand filtration   103
screening plant   137
seawater   43, 161
sedimentation zone   43
serotonin hormones   128
sewage sludges   205 ff.
sewage treatment plants   2
sexual hormones   128
Silent Spring   IX
simultaneous nitrification   139
size-exclusion chromatography   57,
     181, 205
slow sand filtration   107
sludge   3, 15 f., 23. 62, 81 ff.
sludge samples   56
socioeconomic data   61
solid-phase extraction   95, 181, 195
sorption mechanism   208
sorption to activated carbon   210
standard addition   188
standard deviation   12, 191
steroid hormones   144
storage   190
STPs   216
strongly bound residues   58
supernatant water   91
surface waters   24 ff., 63 ff., 93 ff.,
     216
suspended particulate matter   195

t

tandem mass spectrometry 156
TBEP 87
TCPP 86, 206, 217
TDCP 119
terbutylazine 202
tertiary pond 140
testosterone 133
thiocyanatomethylthiobenzothiazole 164
thiophosphates 43, 171
thyroid hormones 128
toxicity indentification evaluation
  197
transformation 1, 209
triazine herbicides 43, 171
tributylin (TBT) 134
trickling filter 136, 138, 145
triclosan 54, 124, 217
tri-*iso*-butylphosphate 76
trimethylsulfonium hydroxide 57
tri-*n*-butylphosphate 76
triphenylphosphate 74

tris-(1,3-dichloro-propyl)
  phosphate 75
tris-(2-chloro-1-methylethyl)
  phosphate 75
tris-(2-chloroethyl) phosphate 75
tris-(butoxyethyl) phosphate 76

u

undergroud passage 103, 107
uterotrophic assay 71
UV irradiation 103
UV-filters 69

v

vitellogenin 153

w

water framework directive 211
waterworks 102f.
Whelk-O column 7

x

xenobiotics IX

Related Titles

H. F. Bender, P. Eisenbarth

Hazardous Chemicals

**Control and Regulation
in the European Market**

2006

ISBN 10: 3-527-31541-1

ISBN 13: 978-3-527-31541-3

T. Reemtsma, M. Jekel (Eds.)

**Organic Pollutants
in the Water Cycle**

**Properties, Occurrence, Analysis
and Environmental Relevance
of Polar Compounds**

2006

ISBN 10: 3-527-31297-8

ISBN 13: 978-3-527-31297-9

H. Parlar, H. Greim (Eds.)

**Essential Air Monitoring
Methods**

**from The MAK-Collection for
Occupational Health and Safety**

2006

ISBN 10: 3-527-31476-8

ISBN 13: 978-3-527-31476-8

J. Angerer, H. Greim (Eds.)

**Essential Biomonitoring
Methods**

**from The MAK-Collection
for Occupational Health and Safety**

2006

ISBN 10: 3-527-31478-4

ISBN 13: 978-3-527-31478-2

T. C. Marrs, B. Ballantyne (Eds.)

**Pesticide Toxicology
and International Regulation**

2004

ISBN 10: 0-471-49644-8

ISBN 13: 978-0-471-49644-1

F. H. Frimmel, G. Abbt-Braun, K. G.
Heumann, B. Hock,
H.-D. Lüdemann, M. Spiteller (Eds.)

**Refractory Organic
Substances
in the Environment**

2002

ISBN 10: 3-527-30173-9

ISBN 13: 978-3-527-30173-7